COLLINS
GEM

BASIC FACTS

BIOLOGY

T A McCahill
BSc DipEd MIBiol

Adviser
K Davison
BSc MSc DipEd MIBiol

Collins
London and Glasgow

First published 1982
Reprinted 1982

© Wm. Collins Sons & Co. Ltd. 1982

ISBN 0 00 458888 6

Phototypeset and illustrated by
Parkway Illustrated Press

Printed in Great Britain by Collins Clear-Type Press

Introduction

Basic Facts is a new generation of illustrated GEM dictionaries in important school subjects. They cover all the important ideas and topics in these subjects up to the level of first examinations.

Bold words in an entry means a word or idea developed further in a separate entry: *italic* words are highlighted for importance.

Tables of important physical facts, units, and nomenclature, together with the major classifications of the animal and plant kingdoms are collected together at the front of the dictionary.

Series adviser: K. Davison, B.Sc., Dip.Ed., M.Sc., M.I. Biol.

Characteristics of living organisms

1. Feeding:	Assimilation of materials for use in metabolism (including respiration), growth and development, and reproduction.
2. Respiration:	Breakdown of high energy chemical bonds to make energy available for many other functions (e.g. metabolism, movement)
3. Excretion:	Elimination of metabolic byproducts and other substances present in excess of requirements.
4. Movement:	Either movement of the whole organism, or of part of it.
5. Sensitivity:	Ability to detect and respond to changes in the environment.
6. Growth:	Increase in size, and often in complexity (development) leading to a fully mature, functional form.
7. Reproduction:	Production of offspring; the ability to self-replicate.

A comparison of these characteristics is given in the next table.

Animals and plants compared

	Animals	Plants
1	Ingestion and breakdown of complex substances into simpler ones.	Assimilation of simple substances, and synthesis of more complex ones.
2	No fundamental difference.	
3	Vital role in eliminating byproducts from the feeding process.	Less vital, since feeding is a synthetic process.
4	Complex movements of whole and parts of organisms.	Slow movement, generally restricted to parts of organisms (e.g. stomata, flowers).
5	Complex responses to changes in many components of stimuli.	Slow responses to long term stimuli.
6	No fundamental difference.	
7	Both varied, but no fundamental differences.	

Animal diets

Food	Essential component	Main use
Protein	nitrogen, and some amino acid molecular structures	growth
Carbohydrate	high energy bonds	energy
Fat	high energy bonds	energy
Vitamins	some parts of molecular structure	e.g. Vit. A: vision Vit. D: bones
Minerals and elements	e.g. sodium iron iodine calcium phosphorus	nerves, muscles oxygen transport thyroid bones enzymes

Plant diets

assimilated chemical	components	product/function
carbon dioxide	carbon, oxygen	carbohydrate
water	hydrogen, oxygen	
nitrates, etc.	nitrogen	+ carbohydrate → protein
sulphates, etc.	sulphur	
other salts and elements	e.g. magnesium, calcium, phosphorus	chlorophyll pigment, cell walls, enzymes

Human digestive enzymes

location	glands	enzyme	substrate	product
mouth	salivary	amylase	starch	maltose
stomach	gastric	pepsin	protein	peptides
		rennin	milk protein	ccagulated milk
duodenum	pancreas	amylase	starch	maltose
		lipase	fats	fatty acids + glycerol
		trypsin	protein + peptides	amino acids
ileum		lactase	lactose	glucose + galactose
		lipase	fats	fatty acids + glycerol
		maltase	maltose	glucose + fructose
		peptidase	peptides	amino acids

The human skeleton

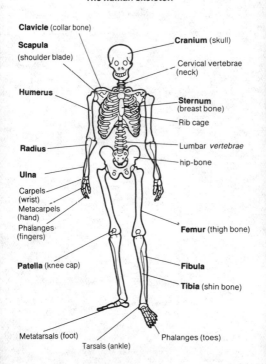

- **Clavicle** (collar bone)
- **Scapula** (shoulder blade)
- **Humerus**
- **Radius**
- **Ulna**
- Carpels (wrist)
- Metacarpels (hand)
- Phalanges (fingers)
- **Patella** (knee cap)
- **Cranium** (skull)
- Cervical vertebrae (neck)
- **Sternum** (breast bone)
- Rib cage
- Lumbar *vertebrae*
- hip-bone
- **Femur** (thigh bone)
- **Fibula**
- **Tibia** (shin bone)
- Metatarsals (foot)
- Tarsals (ankle)
- Phalanges (toes)

The carbon cycle

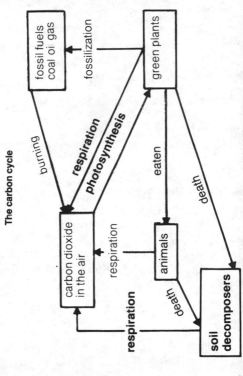

The nitrogen cycle

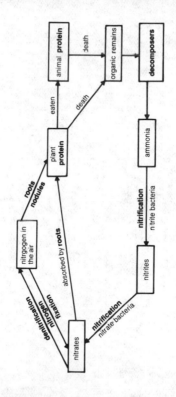

Life history

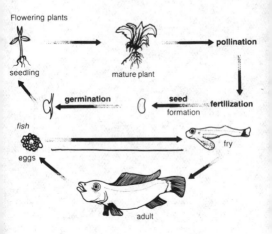

Flowering plants

seedling

mature plant

pollination

germination

seed
formation

fertilization

fish

eggs

fry

adult

The Geological Time-Scale

Era	Period	Beginning (10^6 years ago)
Cenozoic	Quaternary (Q)	1
	Tertiary (Te)	55
Mesozoic	Cretaceous (Cr)	120
	Jurassic (J)	155
	Triassic (Tr)	190
Palaeozoic	Permian (P)	215
	Carboniferous (Car)	300
	Devonian (D)	350
	Silurian (S)	390
	Ordovician (O)	480
	Cambrian (Cam)	550

The occurrence of fossils in relation to this time-scale is
shown in the next table.

The Fossil Record

	Vertebrates	Invertebrates	Plants
Q	Hominids appear	Arthropods, molluscs abundant	
Te	Mammals emerge, dinosaurs extinct	Modern groups emerge	
Cr	Dinosaurs dominant	Ammonoids extinct	Flowering plants emerge
J	Birds emerge	Modern crustacea emerge, ammonoids abundant	
Tr	Reptiles flourish, dinosaurs emerge	Marine forms decline	

P	Amphibia decline	Trilobites extinct	Conifers emerge
Car	Reptiles emerge		
D	Amphibia emerge, fish abundant	Insects emerge	Mosses, horsetails, ferns emerge
S		Trilobites decline, brachiopods abundant	
O	Fish emerge		First land plants emerge
Cam		Most invertebrate phyla present; trilobites and brachiopods flourish	

The Greek Alphabet

Name	Capital	Lower case	Equivalent	
alpha	A	α	a	
beta	B	β	b	
gamma	Γ	γ	g	
delta	Δ	δ	d	
epsilon	E	ϵ	e	(as in *fed*)
zeta	Z	ζ	z	
eta	H	η	e	(as in *feed*)
theta	Θ	φ	th	
iota	I	ι	i	
kappa	K	κ	k	
lambda	Λ	λ	l	
mu	M	μ	m	
nu	N	ν	n	
xi	Ξ	ξ	ks	
omicron	O	o	o	(as in *cot*)
pi	Π	π	p	
rho	P	ρ	r	
sigma	Σ	σ	s	
tau	T	τ	t	
upsilon	Υ	υ	u	
phi	Φ	ϕ	ph	
chi	X	χ	ch	(as in *loch*)
psi	Ψ	ψ	ps	
omega	Ω	ω	o	(as in *coat*)

Chemical Elements

In this Table, we give the names and symbols of the chemical elements, with their proton numbers (Z), numbers of isotopes (n_i) and melting and boiling temperatures (T_m, T_b).

Element	Z	n_i	T_m/°C	T_b/°C
actinium Ac	89	7	1230	3100
aluminium Al	13	7	660	2400
americium Am	95	8	1000	2600
antimony Sb	51	18	631	1440
argon A	18	7	−190	−186
arsenic As	33	11	−	610
astatine At	85	7	250	350
barium Ba	56	16	710	1600
berkelium Bk	97	6	?	?
beryllium Be	4	4	1280	2500
bismuth Bi	83	12	271	1500
boron B	5	4	2030	3700
bromine Br	35	18	−7	58
cadmium Cd	48	18	321	767
caesium Cs	55	15	27	690
calcium Ca	20	11	850	1450
californium Cf	98	7	?	?
carbon C	6	6	3500	3900
cerium Ce	58	13	804	2900
chlorine Cl	17	10	−101	−34
chromium Cr	24	8	1900	2600
cobalt Co	27	10	1490	2900
copper Cu	29	10	1080	2580

Element	Z	n_i	T_m/°C	T_b/°C
curium Cm	96	7	1340	?
dysprosium Dy	66	12	1500	2300
einsteinium Es	99	10	?	?
erbium Er	68	10	1530	2600
europium Eu	63	12	830	1450
fermium Fm	100	7	?	?
fluorine F	9	4	−220	−188
francium Fr	87	5	30	650
gadolinium Gd	64	14	1320	2700
gallium Ga	31	10	30	2250
germanium Ge	32	13	960	2850
gold Au	79	13	1060	2660
hafnium Hf	72	11	2000	5300
helium He	2	3	−	−269
holmium Ho	67	6	1500	2300
hydrogen H	1	3	−259	−253
indium In	49	19	160	2000
Iodine I	53	17	114	183
Iridium Ir	77	10	2440	4550
Iron Fe	26	8	1539	2800
krypton kr	36	19	−157	−153
lanthanum La	57	8	920	3400
lawrencium Lw	103	1	?	?
lead Pb	82	24	327	1750
lithium Li	3	4	180	1330
lutetium Lu	71	5	1700	3300
magnesium Mg	12	6	650	1100
manganese Mn	25	9	1250	2100

Element	Z	n_i	$T_m/°C$	$T_b/°C$
mendelevium Md	101	1	?	?
mercury Hg	80	16	−39	357
molybdenum Mo	42	15	2600	4600
neodymium Nd	60	13	1020	3100
neon Ne	10	7	−250	−246
neptunium Np	93	8	640	3900
nickel Ni	28	11	1450	2800
niobium Nb	41	15	2400	5100
nitrogen N	7	6	−210	−196
nobelium No	102	1	?	?
osmium Os	76	13	3000	4600
oxygen O	8	6	−219	−183
palladium Pd	46	17	1550	3200
phosphorus P	15	7	44	280
platinum Pt	78	12	1770	3800
plutonium Pu	94	11	640	3500
polonium Po	84	12	250	960
potassium K	19	8	63	760
praseodymium Pr	59	8	930	3000
promethium Pm	61	8	1000	1700
protactinium Pa	91	9	1200	4000
radium Ra	88	8	700	1140
radon Rn	86	7	−71	−62
rhenium Re	75	7	3180	5600
rhodium Rh	45	14	1960	3700
rubidium Rb	37	16	39	710
ruthenium Ru	44	12	2300	4100
samarium Sm	62	14	1050	1600

Element	Z	n_i	T_m/°C	T_b/°C
scandium Sc	21	11	1400	2500
selenium Se	34	16	220	690
silicon Si	14	6	1410	2500
silver Ag	47	16	960	2200
sodium Na	11	6	98	880
strontium Sr	38	13	77	1450
sulphur S	16	7	119	445
tantalum Ta	73	11	3000	5500
technetium Tc	43	14	2100	4600
tellurium Te	52	22	450	1000
terbium Tb	65	8	1360	2500
thallium Tl	81	16	300	1460
thorium Th	90	9	1700	4200
thulium Tm	69	10	1600	2100
tin Sn	50	21	231	2600
titanium Ti	22	8	1680	3300
tungsten W	74	10	3380	5500
uranium U	92	12	1130	3800
vanadium V	23	7	1920	3400
xenon Xe	54	22	−111	−108
ytterbium Yb	70	11	820	1500
yttrium Y	39	12	1500	3000
zinc Zn	30	13	420	907
zirconium Zr	40	12	1850	4400

Units used in biology

Length
1 metre (m) = 100 centimetres (cm)
1 centimetre = 10 millimetres (mm)
1 millimetre = 1000 microns (μ)
1 micron = 1000 nanometres (nm)

Volume
1 litre (l) = 1000 cm^3 (millilitres (ml))

Mass
1 tonne = 1000 kilogrammes (kg)
1 kilogramme = 1000 grammes (g)

Temperature
boiling point of water = 100 °Celsius (°C)
freezing point of water = 0°C
normal average human body temperature = 37°C

Energy
1 kilojoule (kJ) = 1000 Joules (J) = 240 calories (C)

Food type	Energy value
Carbohydrate	17 kJ/G
Protein	17 kJ/G
Fat	39 kJ/G

Terms used in biology

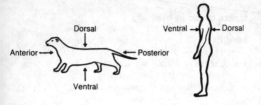

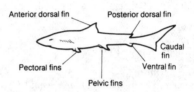

Dorsal Of, on, or, near that surface of an organism which is normally directed upwards, although in humans it is directed backwards. Opposite of ventral.

Ventral Of, on, or, near that surface of an organism which is normally directed downwards, although in humans, it is directed forwards. Opposite of dorsal.

Anterior At, or, near, the leading end of an animal.

Posterior At, or near, the hind end of an animal.

Pectoral Of, on, or near, the chest or breast.

Pelvic Of the pelvis, which is the cavity in the posterior part of the trunk in many vertebrates.

Caudal Of, on, or near, the tail.

Pulmonary Of the **lungs**.

Renal Of the **kidneys**.

Hepatic Of the **liver**.

Gastric Of the **stomach**.

Cardiac Of the **heart**.

Cranial Of the **cranium**.

Neural Of the **nervous system**.

Optic Of the **eye**.

Auditory Of the **ear**.

Costal Of the ribs.

Cervical 1) Of the neck. 2) Of the **cervix**.

Olfactory Of the sense of smell.

Oral Of the mouth.

Sac Pouchlike structure in plants and animals, for example, **stomach**.

Motile Describing organisms or parts of organisms which can move.

Longitudinal section Section through an organism or part of an organism running *length ways*.

Transverse section Section through an organism or part of an organism running *crosswise*.

In vitro Describing biological experiments or observations conducted *outside* an organism, for example, in a test tube.

In vivo Describing biological experiments or observations conducted *within* living organisms.

pH A measure of the degree of *acidity* or *alkalinity* of a **solution**.

pH scale

acid | alkali

```
 ┌──────────────────────────────────────────┐
 0  1  2  3  4  5  6  7  8  9  10 11 12 13 14
                      ↑
                   neutral
```

Tetrapods Vertebrates with *four* limbs. For example, amphibians, reptiles, birds, mammals.

The major groups of living organisms
See **Classification**.
The animal kingdom (Major **phyla**)

1) Invertebrates Animals without a **vertebral column** (backbone).

Phylum Protozoa   Microscopic **unicellular** animals.

Amoeba

Paramecium

Phylum Porifera   Porous animals often occurring in colonies, for example, *sponges*.

bath sponge

Phylum Coelenterata   Tentacle-bearing animals with stinging cells.

Hydra   *Jelly fish*   *Sea anenome*

Phylum Platyhelminthes   Flatworms.

Planaria   *Tapeworm*

Phylum Annelida Segmented worms

Earthworm

Leech

Sandworm

Phylum Mollusca Soft-bodied animals often with shells.

Snail

Clam

Octopus

Phylum Arthropoda Jointed limbs; **exoskeleton**.

Class Insecta (*louse*)

Class Crustacea (*shrimp*)

Class Arachnida (*spider*)

Class Chilopoda (*centipede*)

Phylum Echinodermata Spiny-skinned marine animals.

Starfish

Sea urchin

Brittle star

2) **Vertebrates** (Phylum Chordata) Animals with a vertebral column.

Class Pisces (Fish)
Fins, Scales, Aquatic.

Class Amphibia (amphibians)
moist, scaleless skin, live both
on land and water.

Trout

Toad

Class Reptilia (reptiles)
dry scaly skin.

Class Aves (birds) feathers,
constant temperature.

Turtle

Robin

Class Mammalia (mammals)   Hair; constant temperature; young suckled with milk.

Sub-class Monotremata
Egg-laying.

Sub-class Metatheria
Pouch-bearing.

Duck billed platypus

Kangaroo

Sub-class Eutheria
True **placenta**

Cat

Horse

Man

Summary

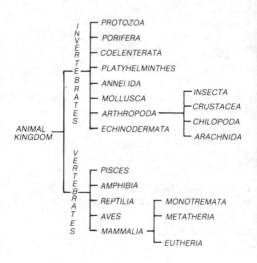

ANIMAL KINGDOM

INVERTEBRATES
- PROTOZOA
- PORIFERA
- COELENTERATA
- PLATYHELMINTHES
- ANNELIDA
- MOLLUSCA
- ARTHROPODA
  - INSECTA
  - CRUSTACEA
  - CHILOPODA
  - ARACHNIDA
- ECHINODERMATA

VERTEBRATES
- PISCES
- AMPHIBIA
- REPTILIA
- AVES
- MAMMALIA
  - MONOTREMATA
  - METATHERIA
  - EUTHERIA

The plant kingdom (Major phyla)

Phylum Thallophyta Unicellular and simple multicellular plants.

9

Class Algae   Photosynthetic; includes unicellular, filamentous, and multicellular types.

Chlamydomonas   Spirogyra   Fucus (seaweed)

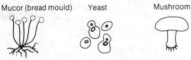

Class Fungi   Heterotrophic; including both **parasites** and **saprophytes**.

Mucor (bread mould)   Yeast   Mushroom

Phylum Bryophyta   Green plants with simple **leaves** and showing **alternation of generations**; moist **habitats**.

Class Hepaticae (liverworts)

Pellia

Class Musci (mosses)

Funaria

Phylum Pteridophyta (ferns; bracken; horsetails)
Green plants, with **roots**, **stems**, **leaves**, and
showing **alternation of generations**.

Fern

Phylum Spermatophyta **Seed** producing plants.

Class Gymnospermae **Seeds** produced in
cones.

Spruce   White pine

Class Angiospermae

Flowering plants; **seeds** enclosed within **fruits**.

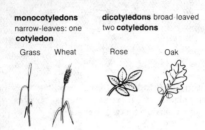

monocotyledons
narrow-leaves: one **cotyledon**

Grass   Wheat

dicotyledons broad-leaved
two **cotyledons**

Rose   Oak

Summary

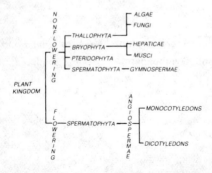

Abdomen In mammals, the part of the body separated from the **thorax** by the **diaphragm**, containing **stomach, liver, intestines**, etc. In insects, the posterior third region of the body.

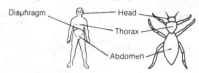

Abscission The shedding of **leaves; fruit**; and unfertilised **flowers** from plants by the formation of a layer of cork **cells** which seal the plant surface and eventually cut off food and water from the part to be shed.

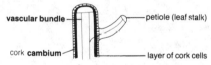

Absorption (of food) The process by which digested food particles pass from the **gut** into the bloodstream. In mammals absorption occurs in the **ileum**. See Ileum.

Accommodation The way in which the **eye** of mammals can change its sharp focus from near to distant objects, and vice-versa, by means of contraction or relaxation of the **ciliary muscles**, so altering the shape and hence the focussing properties of the **lens**. See **Eye**.

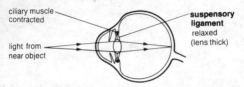

Eye focussed on near object

ciliary muscle contracted

suspensory ligament relaxed (lens thick)

light from near object

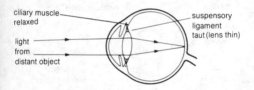

Eye focussed on distant object.

ciliary muscle relaxed

suspensory ligament taut (lens thin)

light from distant object

Active Transport
The movement of materials against a *concentration gradient* using **metabolic energy**.

For example, (1) the uptake of **mineral salts** from **soil** by plant **root hairs**.

For example, (2) the reabsorption of certain substances by the mammalian **kidney**. See **Diffusion**.

A.D.H. (Anti-Diuretic Hormone) A **hormone** secreted in mammals by the **pituitary gland**, which stimulates water reabsorption by the **kidneys**, thus reducing water loss in the **urine**. See **Kidney**.

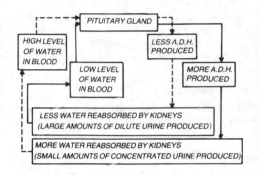

Adipose tissue **Tissue** consisting of **fat** storage **cells**, located in mammals, under the **skin**, around the **kidneys**, etc.

Adrenal glands A pair of **endocrine glands** situated anterior to the mammalian **kidneys** and secreting the **hormone** *adrenaline* which causes increased heartbeat, breathing, etc., in response to conditions of stress. See **Endocrine glands**; **Hormones**.

Aerobe An organism which requires *oxygen* to survive. See **Respiration**.

Alimentary canal The digestive canal in animals. In man it is a tube about nine metres in length running from **mouth** to **anus**. See **Digestion**.

Alleles Alternative forms of an inherited factor (**gene**) which produce different effects for the same inherited feature. For example, in the fruit fly *Drosophila* 'normal wing' and 'vestigial wing' are alleles of the gene controlling wing length. See **Monohybrid inheritance**; **Backcross**; **Incomplete dominance**.

Alternation of generations. In a **life history**, having a generation reproducing by **sexual**

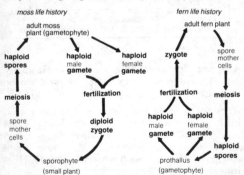

moss life history

adult moss plant (gametophyte)

haploid spores

haploid male gamete

haploid female gamete

meiosis

fertilization

spore mother cells

diploid zygote

sporophyte (small plant)

fern life history

adult fern plant

zygote

spore mother cells

fertilization

meiosis

haploid male gamete

haploid female gamete

prothallus (gametophyte)

haploid spores

reproduction, followed by a generation having **asexual reproduction**, the two generations usually being very different. Alternation of generations is found in some animals, for example, *jelly fish*, and in many plants, for example, *mosses* and *ferns*. See **Diploid**; **Fertilisation**; **Haploid**; **Meiosis**.

Alveoli Air sacs in the mammalian **lung** across which **gas exchange** occurs. See **Gas exchange (Mammals)**.

Amino acids **Organic compounds** which are the sub-units of **proteins**. Altogether some seventy different amino acids are known, but only about twenty to twenty-four are actually found in living organisms, bonded together in chains known as **peptides** which are the basis of protein structure.

Amino Acid Structure

$$NH_2 - \overset{\overset{\displaystyle R}{|}}{\underset{\underset{\displaystyle H}{|}}{C}} - COOH$$

R   variable group (depending on amino acid)

Amino Group   Acid Group

Amnion Fluid-filled sac surrounding and protecting the **embryos** of *mammals*, *birds*, and *reptiles*. See **Pregnancy**.

Amylases **Enzymes** which break down **starch** or **glycogen** into **disaccharides** and **glucose** by **hydrolysis**. For example, *salivary amylase*.

Anabolism See **Metabolism**.

Anaerobe An organism which lives in the absence of *oxygen*. See **Respiration**.

Androecium Collective name for the male reproductive structures of a **flower**, i.e. the **stamens**.

Annual A flowering plant which completes its **life history** from **germination** to death in one season.

Annual ring See **Secondary growth**.

Antagonistic muscles Pairs of **muscles** which produce opposite movements, the contraction of one stimulating the relaxation of the other.

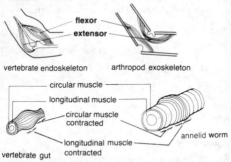

flexor
extensor

vertebrate endoskeleton   arthropod exoskeleton

circular muscle
longitudinal muscle
circular muscle contracted
longitudinal muscle contracted

annelid worm

vertebrate gut

For example, at a **joint**, the contraction of the *flexor* (muscle which bends limb) stimulates the relaxation of the *extensor* (muscle which straightens limb) so that bending occurs. When the joint straightens due to contraction of the extensor, this causes the flexor to relax.

The longitudinal and circular muscles of the vertebrate **gut**, and *annelid* worms also act antagonistically, the former causing **peristalsis**, and the latter movement.

Anther In a **flower**, the upper part of a **stamen** containing **pollen** grains.

Antibiotics Substances formed by certain **bacteria** and *fungi* which inhibit the growth of other **micro-organisms**. For example, *penicillin; streptomycin.*

Antibodies **Proteins** produced by vertebrate **tissues** as a reaction to *antigens* i.e., materials foreign to the organism. (For example, in man, **micro-organisms** and their *toxins*, or transplanted **organs** or tissues). Antibodies react with antigens as shown below.

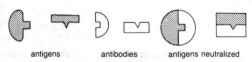

antigens   antibodies   antigens neutralized

Antigens See **Antibodies**.

Anus Terminal opening of the **alimentary canal** in mammals, through which **faeces** are shed. The anal orifice is opened and closed by a **sphincter muscle**.

Aorta The largest **artery** in the **circulatory system** of mammals, which carries **blood** from the left **ventricle** of the **heart** to the rest of the body.

Appendix Small narrow sac at the **caecum** of some mammals. In man it is thought to have no important function.

Aqueous Humour Clear fluid filling the front chamber of the vertebrate **eye** between the **cornea** and the **lens**. See **Eye**.

Arteries Blood vessels which transport **blood** from the **heart** to the **tissues**. In mammals, arteries carry *oxygenated* blood (for an exception to this rule, see **Pulmonary vessels**) and divide into smaller vessels called *arterioles*. They have thick, elastic, muscular walls, in order to withstand the high pressure caused by the **heart beat**.

Section through an artery

thick wall ——— elastic muscle tissue

Asexual reproduction Reproduction in which new organisms are formed from a *single* parent without **gamete** production. The offspring from asexual reproduction are genetically identical to each other and to the parent organism and are referred to as a **clone**. See **Binary fission**; **Budding**; **Spore**; **Vegetative reproduction**.

Assimilation (of food) The process by which digested food particles are incorporated into the **protoplasm** of an organism. For example, in man **glucose** is *oxidised* to produce **energy** via the reaction of **respiration**. Some glucose is converted in **liver** and **muscle cells** into the storage **carbohydrate glycogen** which can be reconverted to glucose if the blood glucose level falls. (See **Insulin**). Excess glucose not stored as glycogen is converted to **fat** and stored in fat storage cells beneath the **skin**, as a long term energy store.

Fatty acids and *glycerol* are reassembled into **fat**. Excess fat is stored as outlined above.

Amino acids are synthesized into **proteins**. Excess amino acids cannot be stored and are disposed of in the liver by **deamination**.

A.T.P. (Adenosine Triphosphate) A chemical compound which acts as a store and a source of **energy** within **cells** A.T.P. is formed from *Adenosine Diphosphate* (A.D.P.) and a *phosphate group* using energy from **respiration**, which can then be released for metabolic processes when A.T.P. is broken down.

For example:

See **Respiration**.

Atrium (Auricle)   See **Heart**; **Heart beat**.

Auditory canal   Tube in the mammalian *outer ear* leading from the **pinna** to the **tympanum**. See **Ear**.

Auditory nerve   A *cranial nerve* in vertebrates conducting **nerve impulses** from the *inner ear* to the **brain**. See **Ear**.

Auricle (Atrium)   See **Heart**; **Heart beat**.

Autoradiograph   A picture obtained when a photographic negative is exposed to living **tissue** into which *radioactive* material has been introduced in order to trace the route of substances through the tissue.

Autotrophic   (Holophytic)   Describing organisms which synthesize complex **organic compounds** from simple non-living **inorganic compounds**. The major autotrophs are *green plants* which use *water* and *carbon dioxide* to make food via **photosynthesis**. For this reason green plants are also called **food producers**. See **Heterotrophic**.

Auxins Plant **hormones** which control many aspects of plant growth, for example, **tropisms**, by stimulating **cell division** and elongation.

Axon See **Nerve cell**; **Synapse**.

Backcross A **genetics** cross in which a **heterozygous** organism is crossed with one of its **homozygous** parents. Thus two backcrosses are possible.

For example, in the fruit fly *Drosophila, normal wing* is **dominant** to *vestigial wing*, and thus, heterozygous flies will have normal wings. The two backcrosses are shown below.

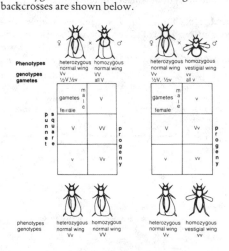

23

The backcross with the **recessive** homozygote is useful in distinguishing between organisms with the same **phenotype** but different **genotypes**. For example, VV and Vv, (such a cross is called a *test cross*).

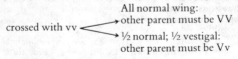

crossed with vv →
→ All normal wing: other parent must be VV
→ ½ normal; ½ vestigal: other parent must be Vv

See **Monohybrid inheritance**.

Bacteria Unicellular organisms with a diameter of 1–2 microns. Some bacteria cause disease, for example, *tetanus* but others are useful, for example, as sources of **antibiotics**.

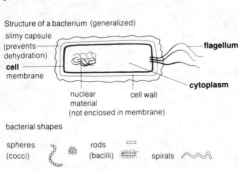

Structure of a bacterium (generalized)

slimy capsule (prevents dehydration)

cell membrane

flagellum

nuclear material (not enclosed in membrane)

cell wall

cytoplasm

bacterial shapes

spheres (cocci)

rods (bacilli)

spirals

Balanced diet The correct nutritional components required for health generally used in reference to human beings and domesticated animals. A balanced diet for humans should contain.

1) Sufficient kilojoules of **energy**
2) **Protein**
3) **Carbohydrate**
4) **Fat**
5) **Vitamins**
6) **Water**
7) **Mineral salts**
8) **Roughage**.

Bile A green alkaline fluid produced in the **liver** of mammals. Bile is stored in the **gall bladder** and is transported via the *bile duct* to the **duodenum** where it causes the *emulsification* of **fat** prior to digestion. See **Digestion**; **Duodenum**.

Binary fission **Asexual reproduction** in **unicellular** organisms in which a single **cell** divides by **mitosis** to produce two cells. Binary fission is common in **bacteria** and *protozoa* such as *Amoeba*, where a single *mother cell* divides into two identical *daughter cells*.

Binary fission in amoeba

nucleus divides **cytoplasm** divides

mother cell 1 2 3 4

5 6 daughter cells

Binomial nomenclature The method of naming organisms devised by *Carl Von Linne (Linnaeus)* in the eighteenth century. Each organism has *two* Latin names, the first, with an initial capital, indicating the **genus**, and the second with a lower case first letter indicating **species**. For example:

Genus	Species	Common name
Canis	familiaris	Domestic dog
Canis	lupus	American wolf

See **Classification**.

Birth (In Humans) The human baby is born as a result of muscular contractions of the **uterus** wall. The *amniotic fluid* escapes, and the baby is pushed through the **cervix** and the **vagina** and thus leaves the mother's body.

The *umbilical cord* is cut, the **placenta** is expelled as *afterbirth* and the baby must now use its own **lungs** for **gas exchange**. See **Fertilization in Man**; **Pregnancy**.

Bladder (Urinary) A sac into which **urine** from the **kidneys** passes via the **ureters**. From the bladder, urine is discharged through the **urethra**. See **Kidney**.

Blind spot See **Eye**.

Blood A fluid found in many animals with the principal function of transporting substances from one part of the body to another. In mammals, blood consists of an aqueous solution called **plasma**, in

26

which there are three types of cells: **platelets, red blood cells; white blood cells**.

The main functions of blood are:

Transport of *oxygen* from **lungs** to **tissues**.

Transport of toxic by-products to the **organs** of **excretion**.

Transport of **hormones** from **endocrine glands** to target organs.

Transport of digested food from the **ileum** to the tissues.

Prevention of infection by **blood clotting; phagocytosis** by white blood cells; **antibody** production.

Blood clotting The conversion of **blood plasma** into a clot, which occurs when blood **platelets** are exposed to air as a result of injury. The platelets produce an **enzyme (thrombin)** which causes the conversion of a soluble **plasma protein**, (**fibrinogen**) into *fibrin*, which forms a meshwork of fibres and the resulting clot restricts blood loss and **micro-organism** entry.

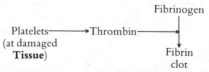

Blood vessels Tubes transporting **blood** around the bodies of many animals, which together with the **heart** make up the **circulatory system**. In vertebrates, the blood vessels consist of **arteries**, **veins**, and **capillaries** which have the following relationship.

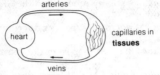

B.M.R. **(Basal Metabolic Rate)** The rate of **metabolism** of a resting animal as measured by oxygen consumption. B.M.R. is the minimum amount of **energy** needed to maintain life and varies with **species**, age and sex.

Bone Tissue in the vertebrate **skeleton** consisting of the **protein collagen** which gives *tensile strength*, and *Calcium phosphate* which gives bone its hardness. Some bones have a hollow cavity containing *bone marrow* in which new **red blood cells** are produced.

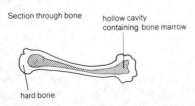

Section through bone   hollow cavity containing bone marrow

hard bone

Bowman's capsule In mammals, a cup-shaped part of a **kidney** tubule or **nephron**. See **Kidney**.

Brain Large mass of **nerve cells** in animals which has a centralized coordinating function. In vertebrates it is found at the anterior end of the body, protected by the **cranium**, and connected to the body, via the **spinal cord** and its *spinal nerves*, or directly by nerves called *cranial nerves*, for example, **optic nerve**; **auditory nerve**.

Section through head to show human brain

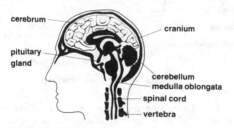

cerebrum

cranium

pituitary gland

cerebellum
medulla oblongata
spinal cord

vertebra

The human brain contains millions of nerve cells which are continually receiving and sending out **nerve impulses**. The remarkable property of the brain is that it translates electrical impulses in such a way that **environmental stimuli** such as *sound* and *light*, are appreciated so that the recipient of the stimuli can respond and adapt to the **environment** in the most appropriate way. The brain also coordinates bodily activities to ensure efficient operation, and stores information so that *behaviour* can be modified as the result of past experience.

Breathing (in Mammals) The inhalation and exhalation of air for the purpose of **gas exchange**. In mammals the gas exchange surface is situated in the **lungs**.

Lungs and associated structures

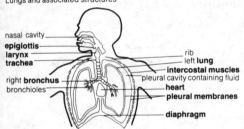

nasal cavity
epiglottis
larynx
trachea
right **bronchus**
bronchioles

rib
left **lung**
intercostal muscles
pleural cavity containing fluid
heart
pleural membranes
diaphragm

The exchange of air in the lungs (*ventilation*) is caused by changes in the volume of the **thorax**, brought about by the action of the **diaphragm** and **intercostal muscles**.

When the diaphragm contracts, it depresses, increasing the volume of the thorax, causing air to rush into the lungs. Relaxation of the diaphragm reduces the volume of the thorax, and causes exhalation of air. The action of the diaphragm is accompanied by the raising and lowering of the *rib cage*, which is necessary to accommodate the changes in lung volume. These rib cage movements are caused by contraction and relaxation of the intercostal muscles.

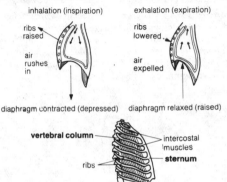

inhalation (inspiration)

ribs raised

air rushes in

diaphragm contracted (depressed)

exhalation (expiration)

ribs lowered

air expelled

diaphragm relaxed (raised)

vertebral column

intercostal muscles

sternum

ribs

Breathing rate The rate of **lung** ventilation. In man **breathing** movements are controlled by the **medulla oblongata** in the **brain**, which is sensitive to the *carbon dioxide* concentration of the **blood**. If the carbon dioxide concentration rises sharply as the result of increased **respiration**, for example, during exercise, the brain sends **nerve impulses** to the **diaphragm** and **intercostal muscles** which react by increasing the rate and depth of breathing. This

31

accelerated breathing rate helps to expel the excess carbon dioxide and increases the supply of oxygen to respiring cells.

Bronchus One of two air passages branching from the **trachea** in lunged vertebrates. See **Lungs**.

Budding Asexual reproduction in which a new organism develops as an outgrowth or *bud* from the parent, the offspring often becoming completely detached from the parent. Budding is common among *Coelenterates*, for example, *Hydra* and **unicellular** fungi, for example, *yeast*.

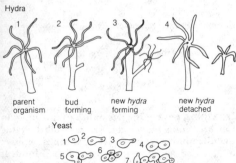

Hydra

| 1 parent organism | 2 bud forming | 3 new *hydra* forming | 4 new *hydra* detached |

Yeast

Bulb **Organ** of **vegetative reproduction** in flowering plants, consisting of a modified **shoot** whose short **stem** is enclosed by fleshy scale-like **leaves**. In the growing season, one or more buds within the bulb develop into new plants using food stored in the bulb. Bulb-producing plants include *tulip*; *daffodil*; *onion*.

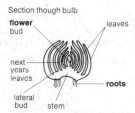

Section though bulb

flower bud

leaves

next years leaves

lateral bud

stem

roots

Caecum   Part of the mammalian **gut** at the entry to the large **intestine**. In man it has no important function, but in **herbivores**, it is an important site of **cellulose digestion**.

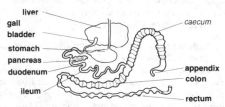

Herbivore digestive system

liver

gall bladder

stomach

pancreas

duodenum

ileum

caecum

appendix

colon

rectum

Cambium   Plant **meristem** within **vascular bundle** which forms new **xylem** and **phloem cells** during **secondary growth**.

Capillaries Blood vessels, formed from *arterioles* and forming a network at vertebrate **tissues**, the **blood** eventually draining into *venules* and then **veins**.

Capillary walls are only one **cell** thick, allowing **diffusion** of substances between the blood and the tissues via a liquid called *tissue fluid* (**lymph**).

Section through a capillary   Capillaries and tissues

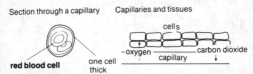

cells

– oxygen ——— carbon dioxide

capillary

red blood cell   one cell thick

Carbohydrates Organic compounds containing the elements Carbon (C), Hydrogen (H) and Oxygen (O) and with the general formula CH_2O. Carbohydrates are either individual *sugar* units or chains of sugar units bonded together.

Importance of Carbohydrates:

1) Simple carbohydrates, particularly **glucose** are the principal **energy** source within cells.
2) Long-chain carbohydrates form some structural cell components, for example, **cellulose** in plant cell walls, and also act as food reserves, for example, **glycogen** in animals; **starch** in plants.

The three main carbohydrate types are **monosaccharides**; **disaccharides** and **polysaccharides**.

Carbon Cycle The circulation of the element *Carbon* and its compounds, in nature, caused mainly by the **metabolic** processes of living organisms. The carbon cycle is summarized below.

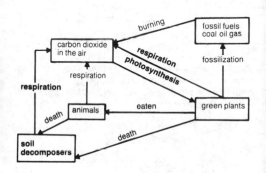

Carnivore An animal that feeds on flesh. Carnivores include dogs, cats, etc., and have a **dentition** adapted for killing prey, shearing raw flesh, and cracking bones. The outstanding features of carnivore dentition are the large piercing *canine* teeth, and the shearing *carnassial* teeth. The lower jaw can usually only move up and down forming an effective clamp on the prey. An example is the dog, shown in the diagram below.

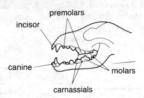

See **Teeth**.

Carpel Female part of a **flower** containing an **ovary** in which there are varying numbers of **ovules** containing **embryo sacs**, within which are the female **gametes**.

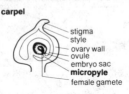

See **Fertilization in plants**.

Cartilage Supporting **tissue** found in vertebrates. In mammals there is cartilage in the **larynx**, **trachea**, **bronchi** and at the ends of **bones** at moveable **joints**, while in some fish, for example, shark, the entire skeleton is cartilage.

Catabolism See **Metabolism**.

Cells Microscopic units making up living organisms. They consist of a mass of **protoplasm** bounded by a *cell wall* (plants) or a *cell membrane* (animals).

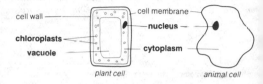

plant cell

animal cell

Nucleus: Contains the hereditary material, the **chromosomes**, and controls the cell's activities.

Cytoplasm: The liquid 'body' of the cell in which the chemical reactions of life occur, for example, **respiration**.

Cell Membrane: Controls the entry and exit of materials, allowing certain substances through, but preventing the passage of others. Such a membrane is described as **selectively permeable**.

Cell Wall: Contains **cellulose** and gives shape and rigidity to plant cells.

Chloroplasts: Structures within green plant cells where **photosynthesis** occurs.

Vacuole: Fluid-filled space making up most of a plant cell's volume. The vacuole exerts pressure on the cell wall, making the plant cells, and hence the plant, firm and resilient.

Cell differentiation Process of change in **cells** during **growth** and development, whereby previously *undifferentiated* cells become specialized for a particular function as a result of structural changes.

For example in plant cells, after **cell division**, the *daughter cells* increase in size (*elongation*), by absorbing water.

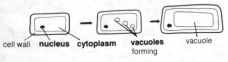

cell wall **nucleus cytoplasm vacuoles** vacuole
forming

After elongation, *cell differentiation* occurs as the result of **protoplasm** and *cell wall* changes. For example:

1) Some cells have their walls strengthened by additional **cellulose**. for example, **cortex**; **epidermis**.
2) Some cells have **lignin** deposited in their walls, for example, **xylem**.
3) Some cells develop extra *organelles*, for example, the high number of **chloroplasts** in **leaf** *palisade mesophyll* cells.

Cell Division Of two kinds:

1) **Mitosis**: by which a **cell** divides into two *daughter cells* with exactly the same number and types of **chromosomes** as the *mother cell*. This is the normal process of **growth** and **tissue** repair,

and of **asexual reproduction** in some simple organisms. See **Mitosis**.

2) **Meiosis** (Reduction Division): By which a cell divides to give *sex cells* or **gametes**, containing half the number of **chromosomes** found in the *mother cell*. This process is part of **sexual reproduction**. See **Meiosis**.

Cellulose Polysaccharide carbohydrate which forms the framework and gives strength to plant **cell** walls. Cellulose remains undigested in the human **gut** but has an important role as **roughage**. In mammalian **herbivores**, at the **caecum**, and **appendix**, **bacterial populations** produce an **enzyme** called *cellulase* which **digests** cellulose into **glucose**.

Central Nervous System (C.N.S.) That part of the vertebrate **nervous system** which has the highest concentration of **neurone** *cell-bodies* and **synapses**, i.e., the **brain** and **spinal cord**.

Cerebellum Region of vertebrate **brain** which in mammals controls *balance* and *muscular coordination* allowing precise controlled movements in activities such as walking and running. See **Brain**.

Cerebrum (cerebral hemispheres) Region of vertebrate **brain** which, in mammals makes up the largest part of the brain. In man the cerebrum consists of right and left hemispheres, the outer part made up of **neurone** *cell-bodies* (*grey matter*), the inner part consists of *nerve fibres* (*white matter*). The human cerebrum is responsible for the higher mental skills such as *memory*; *thought*; *reasoning*; *intelligence*. The cerebrum also contains localized areas concerned with specific functions. Areas receiving **nerve impulses** from **receptors** are called *sensory areas*, while those sending out impulses to **effectors** are called *motor areas*. This localization of function is shown below in the human left cerebral hemisphere.

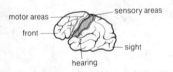

See **Brain**.

Cervix Posterior region of mammalian **uterus**, leading into the **vagina**. See **Fertilisation in Man**.

Chemotropism **Tropism** relative to chemical substances. The growth of **pollen** tubes towards the **ovary** is an example of *positive chemotropism*. See **Fertilization in Plants**.

Chloroplasts Structures in the **cytoplasm** of green plant **cells**, at which **photosynthesis** occurs, chloroplasts contain the green pigment *chlorophyll* which absorbs the *light energy* used in photosynthesis.

chloroplasts in a **palisade mesophyll** cell

chloroplasts in the alga *Spirogyra*

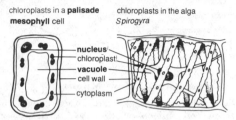

- nucleus
- chloroplast
- **vacuole**
- cell wall
- cytoplasm

Choroid Layer of **cells** outside the **retina** of the vertebrate **eye**. See **Eye**.

Chromosomes The hereditary material within the **nucleus** of **cells**, which links one generation with the next. Each **species** has characteristic numbers and types of chromosomes.

For example, in man, *the chromosome number* is 46. As a result of **cell division**, the chromosome number is maintained during *mitotic growth*. **Haploid gametes** contain half the **diploid** chromosome number and contribute an equal number of chromosomes to the **zygote** at **fertilization**.

Chromosomes control cellular activity, consisting of sub-units called **genes** which contain coded

41

information in the form of the chemical compound **D.N.A.** In diploid cells, chromosomes occur in similar pairs known as *homologous pairs*. Thus a human diploid cell contains twenty-three pairs of **homologous chromosomes**.

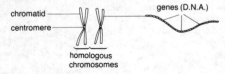

See **D.N.A.**; **Genes**; **Meiosis**; **Mitosis**.

Cilia Microscopic motile threads projecting from certain **cell** surfaces which stroke rhythmically together like oars. Cilia occur in certain vertebrate **epithelia** where they cause movement of particles in **trachea**, **oviduct**, **uterus**, etc. In some protozoa, for example, *Paramecium*, cilia cause movement of the whole organism.

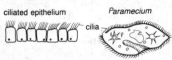

See **Flagellum**.

Ciliary muscle Tissue in the vertebrate **eye** responsible for **accommodation**. See **Eye**; **Accommodation**.

Circulatory system Any system of vessels in animals through which fluids circulate, for example, **blood** circulation; **lymphatic system**.

In mammals there are two overlapping blood circulations, i.e., there is a circulation between **heart** and **lungs** and a circulation between heart and body. This arrangement is called a *double circulatory system*. Blood flows through both circulations, always in the same direction, passing repeatedly through the heart.

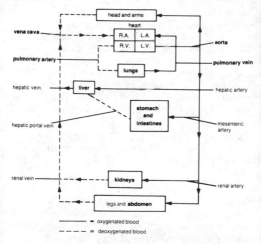

43

Class Unit used in the **classification** of living organisms, consisting of one or more **orders**.

Classification The method of arranging living organisms on the basis of similarity of structure into groups which show how closely they are related to each other and also indicate evolutionary relationships. The modern system of classification was devised by *Carl von Linne* (*Linnaeus*) in the eighteenth century. Organisms are first sorted into large groups called **kingdoms** which are divided into smaller groups called **phyla**, then **classes**, **orders** and **families**, each sub-division producing sub-sets containing fewer and fewer organisms, but with more and more common features. Ultimately organisms are grouped in *genera* (singular **genus**) which are groups of closely related **species**. The classification of some organisms is shown opposite.

Clavicle Ventral **bone** of the shoulder-girdle of many vertebrates articulating with the **scapula** and **sternum**. In man, the 'collar-bone.' See **Endo-skeleton**.

Cloaca Posterior region of the **alimentary canal** in most vertebrates (but excluding mammals) into which the terminal parts of the **intestine**, **kidney** and reproductive ducts open.

	man	sugar maple	dog	white oak
kingdom	animal	plant	animal	plant
phylum	chordata	spermatophyta	chordata	spermatophyta
class	mammalia	angiospermae	mammalia	angiospermae
order	primates	sapindales	carnivora	fagales
family	hominidae	aceraceae	canidae	fagaceae
genus	homo	acer	canis	quercus
species	sapiens	saccharum	familiaris	alba

Clone Group of organisms which are *genetically identical* to each other, having been produced by **asexual reproduction**.

Parent organism $\xrightarrow[\text{reproduction}]{\text{Asexual}}$ Clone (identical offspring)

Cochlea Spiral structure in the mammalian *inner ear*, containing an area called the **organ of corti** in which are located **nerve cell** endings which are sensitive to sound vibrations. See **Ear**.

Collagen Fibous **protein**, which is the principal component of vertebrate **connective tissue**, and an important skeletal substance in higher animals, conferring *tensile strength* to **bones**, **tendons**, and **ligaments**.

Colon Region of the **large intestine** in mammals between the **caecum** and **rectum**, which receives undigested food from the **ileum**. In the colon, much of the water is absorbed from the undigested food, and the semi-solid remains (**faeces**) is passed into the **rectum**. See **Digestion**.

Commensalism A **symbiotic** relationship in which one of the organisms benefits, while the other neither suffers nor benefits. For example, a marine worm lives in a shell with a crab, sharing the crab's food, but giving nothing in return.

Community The **populations** of different **species** living in a particular **habitat** and interacting with each other. For example, a *rockpool* habitat may have a community made up of *crabs*, *worms*, *sponges*, *seaweeds*, etc.

Compensation point (of green plants) The light intensity at which the rate of *carbon dioxide uptake* (**photosynthesis**) is exactly equal to the rate of *carbon dioxide production* (**respiration**). In a single day there are *two* compensation points when the rate of photosynthesis (**carbohydrate** gain) is exactly balanced by the rate of respiration (**carbohydrate** loss). See graph below.

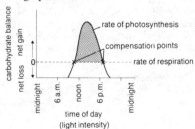

The shaded area of the graph represents the carbohydrate not used in respiration, i.e., **growth**.

Competition The demand within a **community** by organisms of the same **species** (intraspecific competition) or organisms of different species

47

(interspecific competition) for a common resource such as food or light which is in limited supply. Competition often results in the elimination of one organism by another, or even in the elimination of one species, as happens when two species of *Paramecium* compete for food.

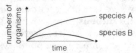

Condensation The synthesis of complex **organic compounds** by **enzyme** action, involving the removal of water from two adjacent molecules. The resulting bond being described as a *condensation bond*. Condensation is the mechanism by which compounds such as **proteins**, **polysaccharides**, and **nucleic acids** are synthesized in living **cells**.

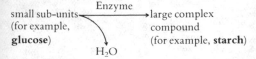

See **Peptide; Hydrolysis**.

Cone 1) Reproductive structure of *gymnosperms*, for example, pines.

2) Light-sensitive **nerve cell** in the **retina** of most vertebrate **eyes**. See **Eye**.

Connective tissue Supporting and packing **tissue** in vertebrates, consisting mainly of **collagen** fibres, in which are embedded more complex structures, such as **blood vessels**; *nerve fibres*, etc.

Contractile vacuole Small sac(s) in the **cytoplasm** of fresh-water *Protista*, the function of which is **osmoregulation**, i.e., in response to water entry by **osmosis**, the vacuole expands as it fills with water, and then *contracts*, discharging its contents out of the **cell**.

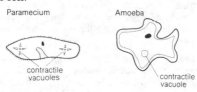

Paramecium

Amoeba

contractile vacuoles

contractile vacuole

Control experiment A test set up in a scientific investigation in which the factor being investigated is kept constant, so that the result of another test in which this factor is varied can be compared. See **Scientific method**.

Copulation The coupling of male and female animals for the purpose of **fertilisation**. In humans the **penis** is inserted in the **vagina** and the **spermatozoa** are released. See **Fertilization in Man**.

Corm Organ of **vegetative reproduction** in flowering plants consisting of an underground **stem**

containing a food store and buds which develop into new plants. Examples of corms include *crocus* and *gladiolus*.

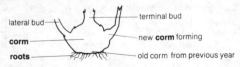

Section through crocus **corm**

lateral bud — terminal bud

corm — new **corm** forming

roots — old corm from previous year

Cornea Transparent **tissue** at the front surface of the vertebrate **eye**, continuous with the **sclerotic** and involved in focussing the image on the **retina**. See **Eye**.

Cortex Animals: The outer layer of an organ, for example, mammalian **kidney**. See **Medulla**.
Plants: Layer of **cells** between the **epidermis** and the **vascular tissue**. Cortex cells are packing and supporting tissue, and in some cases, may store food. See **Leaf; Root; Stem**.

Cotyledon Embryonic **leaf** within a **seed** which supplies food during **germination**, and in some plants is brought above the soil to **photosynthesize** for a time before withering. Flowering plants with *one* cotyledon are called **monocotyledons**, and those with *two* are called **dicotyledons**. See **Seed; Germination**.

Cranium The vertebrate skull which protects the **brain**. See **Endoskeleton**.

Crop rotation The practice of growing a different crop in the same area every year in order to prevent **soil depletion**. Since different plants have different **mineral salt** requirements, changing the crop annually prevents depletion of one particular mineral salt. Another benefit is that since different plants have different **root** lengths, they absorb mineral salts from different **soil** depths. Leguminous plants are often included in rotations, because of the **nitrogen fixation** within their **root nodules**. A typical crop rotation might be wheat/ turnips/ barley/clover/wheat, etc.

Cuticle Non-cellular layer secreted by the **epidermis** of plant *aerial* structures and by many invertebrates. Plant cuticles reduce water loss by **transpiration**, while invertebrate cuticles afford protection against mechanical damage and may also be water retentive/repellant.

Cytoplasm That part of the **protoplasm** of a **cell** bounded by the *cell membrane* but *excluding* the **nucleus**. See **Cells**.

Deamination Removal of the *amino* ($-NH_2$) group from excess **amino acids**. In mammals, this occurs in the **liver**, the amino group automatically changing to the toxic compound *ammonia* (NH_3) which is then converted to **urea** and excreted. The remaining carbon-containing group is converted to useful **carbohydrate**.

Decomposers Heterotrophic organisms which cause the breakdown of dead animals and plants and by so doing, release their constituent compounds which can be used by other living organisms. **Soil** decomposers include **bacteria**, *earthworms*, etc. See **Carbon cycle**; **Nitrogen cycle**.

Denaturation Changes occurring in the structure and functioning of **proteins** (for example, **enzymes**) when subjected to extremes of *temperature* or *pH*.

Dendrite (**Dendron**)   See **Nerve cell**; **Synapse**.

Denitrification The conversion by **soil** *denitrifying bacteria* of *nitrates* into *nitrogen* which can re-enter the atmosphere. see **Nitrogen cycle**.

Dental formula A formula describing the **dentition** of a mammal and expressed by writing the number of **teeth** in the *upper* jaw of *one* side of the **mouth** over the number of teeth in the *lower* jaw on *one* side. Dental formula refers to an *adult* mammal with the correct number of teeth. The total number of teeth is found by doubling the dental formula.

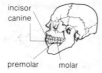

human detition

incisor
canine

premolar   molar

dental formula – incisor ²/₂ canine ¹/₁ premolar ²/₂ molar ³/₃

or $\dfrac{2 \quad 1 \quad 2 \quad 3}{2 \quad 1 \quad 2 \quad 3}$ total number of teeth = 2 × 16 = 32

See **Carnivore**; **Herbivore**.

Dentition The numbers and types of **teeth** in a mammal, described, by a **dental formula**. Dentition reflects an animals diet i.e., an animal has the type of teeth best suited to deal with the type of food on which it feeds. See **Carnivore**; **Herbivore**; **Omnivore**.

Diaphragm Dome-shaped **muscle** separating the **thorax** and **abdomen** in mammals. Contraction and relaxation of the diaphragm is important in **lung** ventilation. See **Breathing**.

Diastema A toothless gap in the **mouth** of many **herbivores**, allowing the tongue to more easily manipulate food.

Diastole See **Heart beat**.

Dicotyledons Larger of the two sub-sets of flowering plants, the other being **monocotyledons**. The characteristics of dicotyledons are:

Two **cotyledons** in the **seed**.

Network of **veins** in **leaves**.

Broad leaves.

Ring of **vascular bundles** in **stem**.

Examples: Hardwood trees; Fruit trees; Herbaceous plants.

Diffusion The movement of particles from a region of high concentration to a region of lower concentration until they are evenly distributed. Diffusion occurs when two different particle concentrations are adjacent.

high particle concentration — low particle concentration

direction of particle movement

The difference in concentration which causes diffusion is called a *concentration gradient*. The greater the concentration gradient, the greater is the rate of diffusion. If no concentration gradient exists, diffusion does not occur, and the situation is described as *equilibrium*.

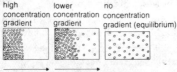

high concentration gradient — lower concentration gradient — no concentration gradient (equilibrium)

rapid diffusion — slower diffusion — no diffusion

The Importance of Diffusion Diffusion is the method by which many substances enter and leave living organisms, and are transported within and between **cells**. For example, (1) uptake of water by plants from **soil**, (2) **gas exchange** between plants and the atmosphere, (3) Gas exchange between **blood** and respiring cells. Where diffusion is too slow for a particular function, substances can be transported more rapidly by **active transport**.

Digestion

The breakdown by **enzyme** action of *large insoluble* food particles into *small soluble* particles, prior to **absorption** and **assimilation**. In many animals, including mammals, digestion and absorption occur in the **alimentary canal**.

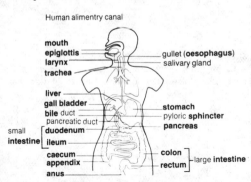

Human alimentry canal

mouth
epiglottis
larynx
trachea

gullet (**oesophagus**)
salivary gland

liver
gall bladder
bile duct
pancreatic duct

stomach
pyloric **sphincter**
pancreas

small intestine {
duodenum
ileum

caecum
appendix
anus

colon
rectum } large **intestine**

55

Diploid Describing a **nucleus; cell** or organism in which the full complement of **chromosomes** is present and occurs as **homologous** pairs. All animal cells, except **gametes**, are diploid, since gametes contain half the diploid number (**haploid**) as the result of **meiosis**. See **Chromosome; Fertilization; Haploid; Meiosis**.

Disaccharides Double sugar **carbohydrates** consisting of *two* **monosaccharides** linked together by **condensation** bonds. For example, maltose is two **glucose** units joined together, while *sucrose* is one *glucose* unit linked with one *fructose* unit.

Disaccharide structure

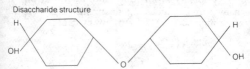

D.N.A. (Deoxyribose Nucleic Acid) **Nucleic acid** which is the major constituent of **genes** and hence **chromosomes**. D.N.A. consists of a double

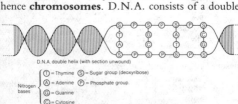

D.N.A. double helix (with section unwound)

Nitrogen bases
- (T) = Thymine
- (A) = Adenine
- (G) = Guanine
- (C) = Cytosine
- (S) = Sugar group (deoxyribose)
- (P) = Phosphate group.

polynucleotide chain, twisted into a *helix*, the two chains being held together by bonds between *nitrogen base pairs*.

The nitrogen bases can only link: (T)—(A); (A)—(T); (G)—(C); (C)—(G) and these are known as complementary pairs. The numbers and sequence of base pairs in the D.N.A. polynucleotide chain represent coded information (*the genetic code*) which acts as a blueprint for the transfer of hereditary information from generation to generation. See **Genes**.

Dominant One of a pair of **alleles** which is always expressed in a **phenotype**, the other being described as **recessive**. See **Monohybrid inheritance**; **Backcross**; **Incomplete dominance**.

Duodenum First part of the mammalian **small intestine** leading from the **stomach** via the pyloric **sphincter**. The duodenum receives *pancreatic juice* from the **pancreas** and **bile** from the **liver**, and is an important digestive site.

The pancreatic juice contains **enzymes** which continue the **digestion** of food arriving from the stomach.

Starch $\xrightarrow{\text{Amylase}}$ Maltose

Protein $\xrightarrow{\text{Trypsin}}$ Peptides $\longrightarrow$ Acids Amino

Fat $\xrightarrow{\text{Lipase}}$ Fatty acids + Glycerol

Bile contains *bile salts* which emulsify fat forming small fat droplets, thus increasing the surface area available for **lipase** action.

From the duodenum, the semi-digested food is forced by **peristalsis** into the **ileum**.

Ear Organ of *hearing* and *balance* in vertebrates. Hearing is a sensation produced by vibrations or sound waves which are converted into **nerve impulses** by the ear and transmitted to the **brain**.

Section through human ear

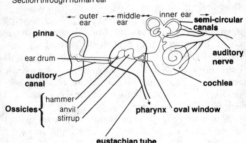

Outer ear **Pinna** is a funnel shaped structure which directs sound waves into the ear and along the **auditory canal**, at the end of which is a very thin membrane, the *eardrum* (**tympanum**), which is made to vibrate by the sound waves.

Middle ear is an air-filled cavity connected to the back of the mouth (**pharynx**) by the **eustachian**

tube, an arrangement which allows air into the middle ear ensuring equal air pressure on both sides of the eardrum.

Within the middle ear, there are three tiny bones, the **ossicles**, named by their shapes: *malleus* (hammer), *incus* (anvil), and *stapes* (stirrup).

The vibrations of the eardrum are transmitted through, and amplified by the ossicles, the stapes finally vibrating against a membrane called the **oval window**, which separates the middle and inner ears.

Inner Ear is fluid-filled and consists of the **cochlea** and **semi-circular canals**.

The vibration of the stapes against the **oval window** sets up waves in the fluid of the cochlea. These waves stimulate **receptor cells** (*hair cells*) causing nerve impulses to be sent via the **auditory nerve** to the brain, where they are interpreted as sounds.

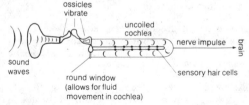

Balance is maintained by the **semi-circular canals** in association with information received from the **eyes** and **muscles**. The semi-circular canals contain

59

fluid and *receptor cells*, which are stimulated by movements of the fluid during changes in posture. The nerve impulses initiated by these cells travel to the brain along the auditory nerve and trigger **responses** which cause the body to maintain normal posture.

Ecdysis See **Exoskeleton**.

Ecosystem A **community** of organisms interacting with each other and with their non-living **environment**. i.e., an ecosystem is a natural unit consisting of living parts (plants and animals) and non-living parts (light, water, air, etc.).

Habitat + *Community* → *Ecosystem*

Ecosystems can be lakes, oceans, forests, etc. The driving force behind all ecosystems is the flow of energy originating from the *sun*.

Effector A specialized animal **tissue** or **organ** that performs a **response** to an **environmental stimulus**, for example, **muscles**; **endocrine glands**. See **Sensitivity**.

Embryo 1) Young animal developed from a **Zygote** as a result of repeated **cell divisions**. In mammals, the embryo develops within the female **uterus**, and in the later stages of **pregnancy**, is called a **foetus**. See **Pregnancy**.

2) Young flowering plant developed from a fertilized **ovum**, which in *seed plants* is enclosed within a **seed** prior to **germination**. See **Seed**.

Embryo sac Structure within the **ovules** of flowering plants in which the female **gametes** are located. See **Carpel**.

Endocrine (ductless) glands Structures which release chemicals called **hormones** directly into the bloodstream in vertebrates and some invertebrates The rate of secretion of hormones is often a response to changes in internal body conditions but may also be a response to **environmental** changes. See **Hormones**.

Major endocrine glands in the human body

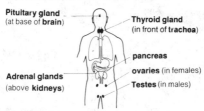

Pituitary gland (at base of **brain**)

Thyroid gland (in front of **trachea**)

pancreas

ovaries (in females)

Adrenal glands (above **kidneys**)

Testes (in males)

Endoskeleton (internal skeleton) Skeleton lying *within* an animal's body, for example, the bony skeleton of vertebrates. Endoskeletons provide *shape*, *support*, and *protection* and in concert with **muscles** produce movement. See **Exoskeleton**.

The main features of the human endoskeleton are shown below

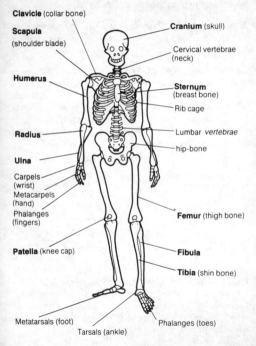

Clavicle (collar bone)

Scapula (shoulder blade)

Humerus

Radius

Ulna

Carpels (wrist)

Metacarpels (hand)

Phalanges (fingers)

Patella (knee cap)

Cranium (skull)

Cervical vertebrae (neck)

Sternum (breast bone)

Rib cage

Lumbar vertebrae

hip-bone

Femur (thigh bone)

Fibula

Tibia (shin bone)

Metatarsals (foot)

Tarsals (ankle)

Phalanges (toes)

Energy The ability to do *work*. In living organisms that work is done in performing the seven *characteristics of life: movement, feeding, reproduction, excretion, growth, sensitivity, respiration.*

Types of energy: heat, light, sound, electrical, chemical, nuclear, potential (stored), *kinetic* (moving). Energy can neither be created nor destroyed, but it can be changed from one form into another. This scientific law is called *the conservation of energy*. Examples:

$$\text{firework:} \quad \text{chemical}\begin{cases}\text{light}\\\text{sound}\\\text{kinetic}\end{cases}$$

$$\text{television:} \quad \text{electrical}\begin{cases}\text{light}\\\text{sound}\end{cases}$$

This concept of energy inter-conversion is important to living organisms, since *green plants* convert the *light energy* of sunlight into the *chemical/ potential energy* of food, via the reaction of **photosynthesis**.

$$\text{sunlight} \xrightarrow[\text{by green plants}]{\text{photosynthesis}} \begin{array}{l}\text{chemical/potential}\\\text{energy of food}\end{array}$$

Other organisms can then release that chemical/ potential energy via the reaction of **respiration** and convert it into other useful forms. For example:

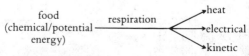

Environment The conditions in which organisms live and which influence the distribution and success of organisms. Many factors contribute to the environment, including: (1) non-living *physical* factors, for example, *temperature, light*, etc. (2) living (*biotic*) factors, for example, **predators, competition**.

The interaction of these factors determines the conditions within **habitats**, and '*selects*' the **communities** of organisms which are best suited to these conditions.

Enzymes **Proteins** which act as *catalysts* within **cells**. Catalysts are substances which cause chemical reactions to proceed, and in cells there may be hundreds of reactions occurring, each one requiring a particular enzyme.

$$A + B \xrightarrow{\quad\times\quad} \text{no reaction}$$

$$A + B \xrightarrow{\text{enzyme}} C + D$$

reactants   products

Enzymes catalyse either *syntheses* by which complex compounds are formed from simple molecules by **condensation**, or *degradations* by which complex molecules are broken down to simple sub-units by **hydrolysis**.

Enzyme characteristics 1) Enzymes are **proteins**.

2) Enzymes work most efficiently within a narrow temperature range. Thus human enzymes work best at 37°C (body temperature) and this is called *optimum temperature*. Above and below this temperature their efficiency decreases, and at temperatures above 45°C most enzymes are destroyed (**denaturation**).

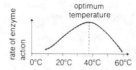

3) Enzymes have an *optimum pH* at which they work most efficiently. For example, the **saliva** enzyme *salivary amylase* works best at neutral or slightly acid pH. The *stomach* enzyme **pepsin** will only function in an acid pH, while the **intestinal** enzyme **trypsin** favours an alkaline pH.

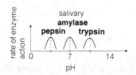

4) The rate of an enzyme-catalysed reaction increases as the enzyme *concentration* increases.

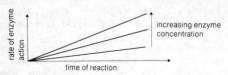

5) The rate of an enzyme–catalysed reaction increases as the *substrate* (the substance on which the enzyme acts) concentration increases, up to a maximum point.

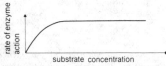

6) Normally an enzyme will catalyse only one particular reaction, a property called *specificity*. For example, the enzyme *catalase* can only degrade the compound *hydrogen peroxide*.

hydrogen peroxide + catalase ⟶ water + oxygen

$$2H_2O_2 \qquad\qquad 2H_2O \quad O_2$$

starch + catalase —✗⟶ No reaction

Naming Enzymes Most enzymes are named by adding the suffix *-ase* to the name of the enzyme's substrate. For example, maltase acts on maltose; urease acts on **urea**, etc.

Enzyme mechanism Enzyme action is explained by the *lock and key hypothesis* in which the enzyme is thought of as a lock into which only certain keys (the substrate molecules) can fit. In this way, the enzyme and the substrate are brought together and the reaction can occur.

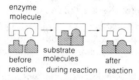

enzyme
molecule

before   substrate   after
reaction   molecules   reaction
                during reaction

The above sequence illustrates an enzyme–catalysed synthesis. By reversing the order of the diagrams they show how an enzyme–catalyse degradation occurs.

Epidermis Protective outermost layer of **cells** in an animal or plant. The epidermis of many **multi-cellular** *invertebrates* is one cell thick and is often covered with a **cuticle**. In most *vertebrates*, the epidermis is the outer layer of **skin**, and in land vertebrates may have several layers of dead cells. In *plants*, the epidermis is one cell thick, and on *aerial* structures, may have a cuticle. See **Leaf**; **Root**; **Stem**.

Epiglottis   Flap of **cartilage** on the ventral wall of the mammalian **mouth**, which closes over during swallowing. See **Digestion**.

Epithelium   Lining **tissue** in vertebrates consisting of closely packed layers of **cells**, covering internal and external surfaces. For example, the **skin** and the lining of the *breathing*, *digestive* and *urinogenital* **organs**. Epithelia may also contain specialized structures, for example, **cilia; goblet cells**.

Erythrocyte   See **Red blood cell**.

Eustachian tube   Tube connecting the *middle ear* to the **pharynx** in tetrapods, and important in equalizing air pressure on either side of the *ear-drum*. See **Ear**.

Evolution   The development of complex organisms from simpler ancestors occurring over successive generations. The most commonly accepted theory is based on *Charles Darwin's* idea of **natural selection**.

Excretion   Elimination of the waste products of **metabolism** by living organisms. The main excretory products are *water*, *carbon dioxide*, and *nitrogenous compounds*, for example, **urea**. In simple organisms excretion occurs through the **cell** *membrane* or **epidermis**, in higher plants via the **leaves**, while most animals have specialized excretory **organs**. For example, in man the **lungs** excrete water and carbon dioxide, and the **kidneys** excrete urea. See **Kidney**.

Exoskeleton (External Skeleton) Skeleton lying *outside* the body of some invertebrates, for example, the cuticle of insects and the *shells* of crabs. Some organisms shed and renew their exoskeletons periodically to allow **growth**, a process known as *moulting* or *ecoysis*.

See **Endoskeleton**.

Eye A **sense organ** responding to light, ranging from very simple structures in invertebrates to the complex organs of insects (*compound eye*) and vertebrates.

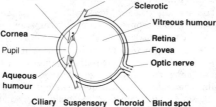

Section through mammalian eye

Iris Eye **muscle**
Sclerotic
Vitreous humour
Cornea
Pupil
Retina
Fovea
Optic nerve
Aqueous humour
Ciliary Suspensory Choroid Blind spot
muscle ligament

Eye muscles enable the eye to move up and down and from side to side.

Sclerotic is a tough protective layer which at the front of the eye forms the transparent **cornea**.

Choroid is a black-pigmented layer under the sclerotic, rich in **blood vessels** supplying food and oxygen to the eye.

Retina is a layer of **nerve cells** which are sensitive to light. There are two types of cells in the retina, named by their shape:

1) **Rods** are very sensitive to low intensity light and are particularly concentrated in the eyes of *nocturnal* animals.

2) **Cones** are sensitive to bright light. There are different types which are stimulated by different wavelengths of light and are thus responsible for *colour vision*. Animals whose retinas lack cones are colour blind, while human colour blindness is caused by a defect in the cones.

The **fovea** (*yellow spot*) is a small area of the retina containing only cones in great concentration and giving the greatest degree of detail and colour.

The **blind spot** is that part of the retina at which *nerve fibres* connected to the rods and cones leave the eye to enter the **optic nerve** which leads to the **brain**. Since there ae no light-sensitive cells at this point, an image formed at the blind spot is not registered by the brain..

The **lens** is a transparent *biconvex* structure which can change curvature and is mainly responsible for focussing light on the retina.

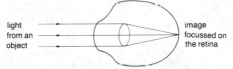

The lens is held in place by **suspensory ligaments** which are attached to the **ciliary muscles**, the contraction or relaxation of which alters the shape of the lens allowing both near and distant objects to be focussed sharply. This is called **accommodation**.

The **iris** is the coloured part of the eye, containing **muscles** which vary the size of the *pupil*, the hole through which light enters the eye. In poor light, the pupils are wide open (*dilated*), to increase the brightness of the image. In bright light the pupils are *contracted* to protect the retina from possible damage. This mechanism is an example of a **reflex action**.

pupil dilation
(dark adapted)

pupil contraction
(light adapted)

Because the pupil is small, light rays enter the eye in such a way that the image at the retina is upside down (inverted). This inversion of the image is corrected by the brain.

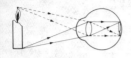

the image is smaller than the object and inverted

The **aqueous** and **vitreous humours** are fluid-filled chambers which maintain the shape of the eye, help in focussing light, and allow nutrients, oxygen, and wastes to diffuse to and from eye cells.

F_1 (generation) (first filial generation) The first generation of **progeny** obtained in breeding experiments. Successive generations are called F_2, etc. See **Monohybrid inheritance**.

Faeces In vertebrates, the solid or semi-solid remains of undigested food, **bacteria**, etc, which is formed in the **colon** and expelled via the **anus**.

Family Unit used in the **classification** of living organisms consisting of one or more *genera* (singular **genus**).

Fats (Lipids) Organic compounds containing the elements *Carbon, Hydrogen, Oxygen*. Fats consist of three *fatty acid* molecules (which may be the same or different) bonded to one *glycerol* molecule.

Fat deposits under the **skin** act as a long term **energy** store, yielding 39 kJ/g when respired. These deposits also provide heat insulation.

Fat is an important constituent of the **cell** membrane and its insolubility in water is utilized in the waterproofing systems of many organisms.

Femur 1) Part of insect limb nearest to the body.

2) The *thighbone* of tetrapod vertebrates. See **Endoskeleton**.

Fermentation The degradation of **organic compounds** in the absence of oxygen for the purpose of **energy** production, by certain organisms, particularly **bacteria** and *yeasts*. Fermentation is a form of anaerobic **respiration**. For example, fermentation by yeast (this reaction is the basis of *brewing*).

$$\text{glucose} \xrightarrow{\text{yeast}} \text{ethanol} + \text{carbon dioxide}$$
$$C_6H_{12}O_6 \qquad 2C_2H_5OH \qquad 2CO_2$$
$$\text{A.D.P.} \quad \textbf{A.T.P.}$$

Fertilization The fusing of **haploid gametes** during **sexual reproduction** resulting in a single cell, the **zygote**, containing the **diploid** number of chromosomes. *External fertilization* occurs when the gametes are passed out of the parents and fertilization and development take place independently of the parents. External fertilization is common in aquatic organisms where the movement of water helps the gametes to meet. Examples.

Fish (for example stickleback)   Amphibians (for example toads)

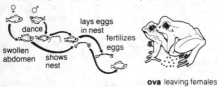

dance
lays eggs in nest
fertilizes eggs
swollen abdomen
shows nest

ova leaving females body

Internal Fertilization is particularly associated with terrestrial animals, for example, insects, birds and mammals, and involves the union of the gametes within the female's body. The advantages of internal fertilization are: (1) the sperms are not exposed to unfavourable dry conditions, (2) the chances of fertilization occurring are increased, (3) the fertilized **ovum** is protected within a shell (birds) or within the female body (mammals).

Examples.

Locusts

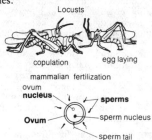

copulation   egg laying

mammalian fertilization

Ovum

Fertilization in man

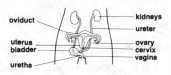

oviduct — kidneys
— ureter
uterus — ovary
bladder — cervix
uretha — vagina

Male reproductive organs

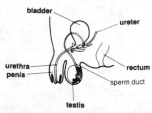

bladder
— ureter

urethra
penis
— rectum
sperm duct

testis

Sperm cells, produced in the testes, are passed out of the penis during **copulation** which involves the penis being inserted in the vagina. The sperms move through the uterus and if an ovum is present in an oviduct, fertilization can occur there. The fertilized ovum (**zygote**) continues moving towards the uterus, dividing repeatedly as it does so. On arrival at the uterus, the zygote, by now a ball of cells, becomes embedded in the prepared wall of the uterus. This is called **implantation** and further development of the **embryo** occurs in the uterus. See **Pregnancy**; **Birth**.

Fertilization in plants In flowering plants, after **pollination**, **pollen** grains deposited on *stigmas* absorb nutrients and *pollen tubes* grow down through the *style* and enter the ovules through the **micropyles**. The tip of each pollen tube breaks down and the male **gamete** enters the ovule and fuses with the female gamete,

After fertilization, the ovule, containing the plant embryo, develops into a **seed**, and the **ovary** develops into a **fruit**. See **Flower**; **Pollen**; **Pollination**.

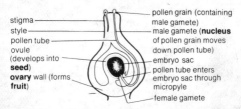

stigma —
style —
pollen tube —
ovule (develops into **seed**) —
ovary wall (forms **fruit**) —

pollen grain (containing male gamete)
male gamete (**nucleus** of pollen grain moves down pollen tube)
embryo sac
pollen tube enters embryo sac through micropyle
female gamete

Fertilizer Substance added to **soil** to increase the quantity or quality of plant growth. When crops are harvested, the natural circulation of soil **mineral salts** is disturbed, i.e., mineral salts absorbed by plants are not returned to soil. This is called **soil depletion** and may render the soil infertile. Fertilizers replenish the soil and are of two types:

1) **Organic** fertilizers such as *sewage*.

2) **Inorganic** fertilizers such as *ammonium sulphate*.

Fibrinogen Soluble **plasma protein** involved with **blood clotting**.

Fibula The posterior of two **bones** in the lower hind-limb of tetrapods. In man the outer bone of the leg below the knee. See **Endoskeleton**.

Flagellum Microscopic motile thread projecting from certain **cell** surfaces and causing movement by lashing back and forth. Flagella are relatively larger than **cilia**, and less numerous, and are responsible for locomotion in many **unicellular** organisms and reproductive cells.

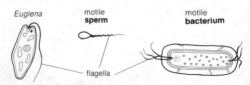

Euglena   motile **sperm**   motile **bacterium**

flagella

77

Flower Organ of **sexual reproduction** in flowering plants (*angiosperms*).

Structure

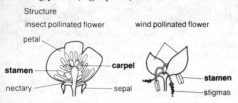

insect pollinated flower   wind pollinated flower

petal

stamen   carpel

nectary   sepal

stamen

stigmas

Insect Pollinated Flowers have brightly coloured and scented petals, and usually have a nectary. The stamens and carpels (with sticky *stigmas*) are within the flower. These adaptations favour *insect pollination*.

Wind Pollinated Flowers are small, often green and unscented, and do not have nectaries. The anthers and feathery stigmas dangle out of the flower thus facilitating *wind pollination*. See **Fertilisation in plants; Pollen; Pollination.**

Foetus Mammalian **embryo** after development of main features. In man this is after about three months of **pregnancy**. See **Pregnancy.**

Food capture Many **heterotrophs** have developed very specialized methods and structures for obtaining food; a variety of examples is given below.

Mammals without teeth *Ant-eaters* have a long sticky tongue for catching ants. *Blue whales* have modified mouth parts to filter **plankton** out of water.

Filter feeding Like the blue whale, many aquatic organisms filter plankton. For example, *Mytilus* (the edible mussel) shown below with one shell removed.

Feeding by sucking *Houseflies* pass **saliva** out onto its food, for example, sugar. **Digestion** begins immediately and the resulting liquid is then taken in by a sucking pad called a *proboscis*.

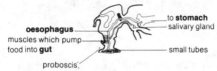

Female *mosquitoes* pierce the human skin, inject a fluid which prevents **blood clotting**, and then suck

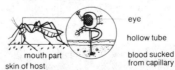

up some blood. This feeding mechanism can cause the disease *malaria*, since the **parasite** involved may be transmitted during feeding.

Butterflies feed on the nectar produced by flowering plants. They suck the nectar from the **flower** by means of a long tube-like *proboscis*, which remains coiled when not in use.

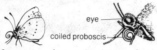

eye

coiled proboscis

Greenflies use a long piercing *proboscis* to suck plant juices from **leaves** and **stems**.

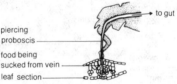

to gut

piercing proboscis

food being sucked from vein

leaf section

Biting without teeth *Locusts* eat their own weight of plant material every day. They have powerful biting jaws called *mandibles* which have very hard biting edges, which are brought together during feeding in a precise and efficient shearing action.

mandibles

Food chain (food web) A food relationship in which **energy** and *carbon* compounds obtained by green plants via **photosynthesis** are passed to other living organisms, i.e., plants are eaten by animals which in turn are eaten by other animals and so on.

green⟶ **herbivore** ⟶ small⟶ large
plant (primary **carnivore** carnivore
(producer) consumer) (secondary (tertiary
consumer) consumer)

The arrows indicate 'is eaten by'. An example of such a food chain is:

plants → insects → lizards → snakes

Not all food chains are as long as the above, for example:

grass → sheep → man
grass → antelope → lion

Such simple food chains seldom exist independently, more often several food chains are linked in a more complicated relationship called a *food web*. The diagram below shows part of the food web in a fresh water pond.

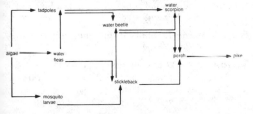

_ All food webs are delicately balanced. Should one link in the web be destroyed, all the other organisms will be affected. For example, in the pond food web, if the *perch* disappeared as the result of disease, the *pike* population would *decrease* while the *water scorpions* would *increase*.

Food consumers Heterotrophic organisms which, after green plants, occupy the subsequent links in a **food chain**.

Food producers Autotrophic organisms mainly *green plants* which occupy the first level in a **food chain**.

Food tests Chemical tests used to identify the components of a food sample. Some common food tests are shown below.

Protein + Biuret reagent → violet/purple colour
(blue)

Reducing + Benedict's reagent $\overset{\text{heat}}{\rightarrow}$ green/yellow/bric
sugar   (blue)   red colour

Starch + Iodine solution → blue/black colour
(Brown)

Fat + Water + Ethanol → white emulsion
(clear)

Vitamin C + Dichlorophenolindo- → D.C.P.I.P.
phenol (D.C.P.I.P.)   (clear)
(blue)

Fovea Area of the **retina** in some vertebrate **eyes**, specialized for acute vision. See **Eye**.

Fruit Ripened **ovary** of a **flower**, enclosing **seeds**, formed as the result of **pollination** and **fertilization**.

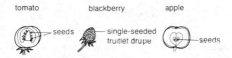

tomato   blackberry   apple

seeds — single-seeded fruitlet drupe — seeds

Fruit and seed dispersal The methods by which most flowering plants spread **seeds** far away from the parent plant, thus: (1) avoiding **competition** for resources, (2) ensuring a wide *colonization*, so that suitable **habitats** are likely to be encountered by a proportion of seeds.

Wind Dispersal Air currents carry the fruits or seeds which usually show an adaptation to increase surface area.

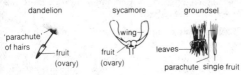

dandelion   sycamore   groundsel

'parachute' of hairs — fruit (ovary)

wing — fruit (ovary)

leaves — parachute single fruit

Animal dispersal *Hooked fruits*, for example, *burdock*, stick to animals coats and may be brushed off some distance from the parent plant. *Succulent fruits*, for example, *strawberry*, are eaten by animals, and the small hard seeds pass through the **gut** unharmed before being released in the *faeces*.

burdock

hooks

strawberry

fruits

Explosive dispersal Unequal drying of part of a fruit causes the fruit to burst, so scattering the seeds.

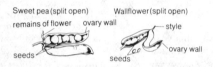

Sweet pea (split open)

remains of flower   ovary wall

seeds

Wallflower (split open)

style

ovary wall

seeds

F.S.H. (Follicle–Stimulating Hormone) Hormone secreted by the vertebrate **pituitary gland**. See **Hormones; Ovulation**.

Gall bladder Small bladder in or near the vertebrate **liver**, in which **bile** is stored. When food enters the **intestine**, the gall bladder empties bile into the **duodenum** via the *bile duct*. See **Digestion**.

Gamete Reproductive **cell** formed by **meiosis** containing *half* the normal **chromosome** number (**haploid**). Human male gametes are **spermatozoa** and the female gametes are **ova** (*egg cells*), which fuse during **fertilization** forming, a **zygote** in which the normal chromosome number (**diploid**) is restored. Thus in Man:

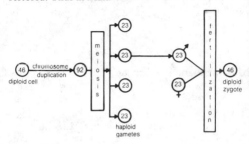

Gas exchange The process by which organisms exchange gases with the **environment** for the purpose of **metabolism**. Most organisms require a continuous supply of the gas *oxygen* for the reaction of **respiration**:

glucose + oxygen → **energy** + carbon + water
                                      dioxide

In addition, *green plants*, require *carbon dioxide* for the reaction of **photosynthesis**:

$$\text{carbon dioxide} + \text{water} \xrightarrow[\text{chlorophyll}]{\substack{\text{light} \\ \text{energy}}} \textbf{carbohydrate} + \text{oxygen}$$

Both reactions use and produce gases which are interchanged between the atmosphere (land organisms) or water (aquatic organisms)

Gas exchange surfaces

Gas exchange takes place across surfaces which have the following characteristics:

1) A *large surface area* for maximum gas exchange.
2) The surface is *thin* to allow easy **diffusion.**
3) The surface is *moist* since gas exchange occurs in **solution.**
4) In animals, the surface has a *good blood supply*, since the gases involved are transported via the **blood.**

Gas exchange (fish)

Gas exchange occurs across **gills** which consist of *gill arches* to which are attached numerous gill filaments. Water is taken in

Position of gills
(gill cover removed)

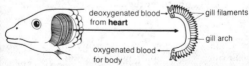

via the mouth and passed over the gills where oxygen dissolved in the water is absorbed into **blood capillaries** while carbon dioxide diffuses into the water.

Gas exchange (insects) Air enters insects through pores called **spiracles** and is carried through a branching system of **tracheae** and thus into smaller branches called *tracheoles* which are in contact with the **tissues**. **Gas exchange** occurs via the fluid in the tracheoles.

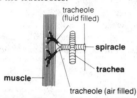

tracheole
(fluid filled)

spiracle

trachea

muscle

tracheole (air filled)

Gas exchange (mammals) **Gas exchange** occurs across the **alveoli** in the **lungs** as the result of

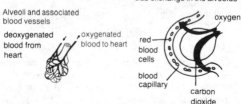

Gas exchange in the alveolus

Alveoli and associated blood vessels

deoxygenated blood from heart

oxygenated blood to heart

oxygen

red blood cells

blood capillary

carbon dioxide

See **Breathing in mammals; Lungs**.

concentration gradients existing between the air in the alveoli and the *deoxygenated blood* ariving from the **heart**. These gradients cause **diffusion** of *oxygen* from the alveoli into **red blood cells** and diffusion of *carbon dioxide* from the **blood** into the alveoli.

Gas exchange (plants)

Terrestrial plants Gas exchange in **leaves** and young **stems** occurs through pores in the **epidermis** called **stomata**. In young **roots** it occurs by diffusion between the roots and air in the **soil**. In older stems and roots where *bark* has formed, gas exchange takes place through gaps in the bark called **lenticels**.

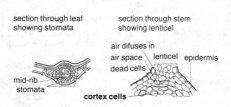

section through leaf
showing stomata

section through stem
showing lenticel

air difuses in
air space   lenticel   epidermis
dead cells

mid-rib
stomata

cortex cells

Aquatic plants Submerged plants, for example, *pondweed* have no stomata, gas exchange occurring by diffusion across the *cell membranes*. Aquatic plants with floating leaves, for example, *water lily*, have stomata only on the upper leaf surface.

Non-green plants *Mushrooms*, for example, do not **photosynthesize** but do carry out **respiration**. Gas exchange occurs by diffusion between the plant cells and the surrounding air.

Genes The sub-units of **chromosomes** consisting of lengths of **D.N.A.** which control the hereditary characteristics of organisms. Genes consist of up to one thousand *base pairs* in a D.N.A. molecule, the particular sequence of which represents coded information. This is known as the *genetic code* and determines the types of **proteins** synthesized by **cells**, particularly enzymes, which then dictate the structure and function of cells and **tissues**, and ultimately organisms, i.e., a cell or an organism is an expresson of the genes it has inherited (and the **environment** in which it lives).

The genetic code is the arrangement of *nitrogen base pairs* in D.N.A. Each group of *three* adjacent base pairs (*triplets*) is responsible for linking together, within the cell, **amino acids** to form protein. The sequence, types and numbers of amino acids determine the nature of the proteins, which in turn determine the characteristics of cells.

For example, the base triplet GTA codes for the amino acid *histidine* while GTT codes for *glutamine*.

Consider two fruit flies (*Drosophila*), one with a gene X controlling body colour, while the other's body colour is controlled by gene Y.

gene X $\xrightarrow[\text{synthesizes}]{}$ enzyme X $\xrightarrow[\text{catalyses}]{}$ pigment X

Gene Y $\xrightarrow{}$ enzyme Y $\xrightarrow{}$ pigment Y

pigment X $\xrightarrow[\text{produces}]{}$ light body

pigment Y $\xrightarrow{}$ dark body

See **Chromosomes**; **D.N.A.**

Genetics The study of *heredity*, which is the transmission of characteristics from parents to offspring via the **genes** in the **chromosomes**. Heredity is investigated by performing *breeding experiments* and then comparing the characteristics of the parents and offspring. Such experiments were first done by *Gregor Mendel* in the 1860s using pea plants. See **Monohybrid inheritance**; **Backcross**; **Incomplete dominance**.

Genotype The genetic composition of an organism, i.e., the particular set of **alleles** in each cell. In breeding experiments, genotypes are represented by symbols, capital letters denoting the **dominant alleles** and small letters denoting the **recessive alleles**. See **Monohybrid inheritance**.

Genus Unit used in the **classification** of living organisms consisting of a number of similar **species**.

Geotropism **Tropism** relative to gravity. Plant **shoots** grow away from gravity (*negative geotropism*), but most *roots* are *positively geotropic*.

Germination The beginning of **growth** in **spores** and **seeds**, which often follows a period of *dormancy*, and which normally proceeds only under certain environmental conditions. For example, the availability of *water* and *oxygen*, and a *favourable temperature*. If these conditions are not present, spores and seeds may remain alive for some time before germinating. In this state, they are said to be *dormant*.

Seed germination in flowering plants Water is absorbed via the **micropyle**.

The **radicle** grows out of the **testa** into the **soil**, and **root hairs** develop.

The **plumule** is pushed up out of the soil, and develops into the first **leaves**, making the seedling independent of its **cotyledons**, and growth continues fuelled by **photosynthesis**.

Germination in the french bean

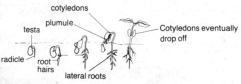

cotyledons

plumule

testa

radicle

root hairs

lateral roots

Cotyledons eventually drop off

Gestation period See **Pregnancy**.

Gills **Gas exchange** surface of aquatic animals. In fish gills are usually *internal*, projecting from the **pharynx**, while in amphibian **larvae**, they are *external*. See **Gas exchange (fish)**.

Glomerulus Knot of **blood capillaries** within the **Bowman's capsule** of the mammalian **kidney**. See **Kidney**.

Glottis The opening of the **larynx** into the **pharynx** of vertebrates.

Glucose Monosaccharide carbohydrate, synthesized in green plants, during **photosynthesis** and serving as an important **energy** source in animal and plant **cells**. See **Monosaccharide**; **Respiration**.

Glycogen Polysaccharide carbohydrate consisting of branched chains of **glucose** units, and important as an **energy** store in animals. In vertebrates, glycogen is stored in **muscle** and **liver cells** and is readily converted to glucose by **amylase enzymes**. See **Polysaccharides**; **Insulin**.

Goblet cells Specialized **cells** in certain **epithelia**, which synthesize and secrete **mucus**. Goblet cells are common in vertebrates, for example, in the **intestinal** and respiratory tracts of mammals.

epthelial cells

mucus secreting goblet cells

Gonads Organs in animals which produce **gametes** and in some cases **hormones**, for example, ovaries and testes.

Graafian follicle Fluid-filled cavity in the mammalian **ovary** within which the **ovum** develops until **ovulation**.

Growth Increase in size and complexity of an organism during development from **embryo** to maturity, resulting from **cell division**; *cell enlargement*, and **cell differentiation**. In plants, growth occurs at certain localized areas called **meristems**, while animal growth goes on all over the body. See **Primary growth**; **Secondary growth**.

Guard cells Paired **cells** bordering **stomata** and controlling the opening and closing of the stomata. The **diffusion** of water into guard cells from adjacent **epidermis** cells causes the guard cells to expand and increase their **turgor**. However they do not expand uniformly, the *thicker, inelastic* cell walls causing them to bend so that the guard cells draw apart forming a stoma. Diffusion of water from guard cells reverses the process and closes the stoma. Stomata are normally open during the day and closed at night.

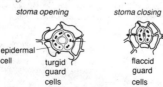

stoma opening   stoma closing

epidermal cell   turgid guard cells   flaccid guard cells

Gut All or part of the **alimentary canal**.

Gynaecium Collective name for the female reproductive structures of a **flower**, i.e., the **carpels**.

Habitat The place where an animal or plant lives, the organism being adapted to the particular **environmental** conditions within the habitat, which may be a *sea-shore*; *pond*; *rockpool*; etc.

Haemoglobin Red pigment containing *iron*, within vertebrate **red blood cells**, responsible for the transport of oxygen throughout the body.

Haemolysis The loss of **haemoglobin** from **red blood cells** as a result of damage to the *cell membrane*. This can be caused by several factors

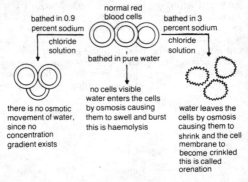

bathed in 0.9 percent sodium chloride solution

normal red blood cells

bathed in 3 percent sodium chloride solution

bathed in pure water

there is no osmotic movement of water, since no concentration gradient exists

no cells visible water enters the cells by osmosis causing them to swell and burst this is haemolysis

water leaves the cells by osmosis causing them to shrink and the cell membrane to become crinkled this is called orenation including **osmosis**, which can be investigated with human red blood cells which have a *solute* concentration equivalent to a 0.9 per cent *sodium chloride* solution.

Haploid Describing a **nucleus**, **cell** or organism having a single set of unpaired **chromosomes**. The haploid number is found in plant and animal **gametes** as the result of **meiosis**. In plants with **alternation of generations**, the **spores** and *gametophyte generation* (gamete producing stage) are haploid.

See **Alternation of generations; Chromosome; Diploid; Meiosis.**

Heart A muscular pumping **organ** which maintains **blood** circulation, and is usually equipped with **valves** to prevent backward flow. In mammals, the

Section through mammalian heart

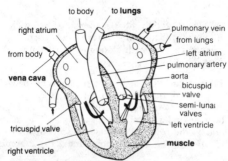

to body · to lungs

right atrium

pulmonary vein from lungs

left atrium

from body

pulmonary artery

vena cava

aorta

bicuspid valve

semi-lunar valves

left ventricle

tricuspid valve

muscle

right ventricle

heart has four chambers, consisting of two relatively thin-walled **atria** (or **auricles**) which receive blood, and two thicker-walled **ventricles** which pump blood out.

The right side of the heart deals only with *deoxygenated blood*, and the left side only with *oxygenated blood*. The wall of the left ventricle is thicker and more powerful than that of the right, since it pumps all round the body, while the right ventricle pumps only to the lungs. See **Circulatory system**; **Heart beat**.

Heart beat The alternative contraction and relaxation of the **heart**. In mammals it consists of two phases:

1) *Diastole* The *atria* and *ventricles* relax, allowing **blood** to flow into the ventricles from the atria.

2) *Systole* The ventricles contract, forcing blood into the **pulmonary artery** and **aorta**. The relaxed atria fill with blood in preparation for the next beat.

Heart beat is initiated by a structure in the *right atrium* called the *pacemaker*, although the *rate* is controlled by the **medulla oblongata** of the **brain** which detects any increase in *carbon dioxide* in the blood as the result of increased **respiration** and is also affected by certain **hormones**, for example,

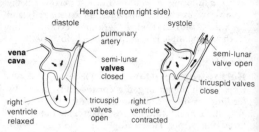

Heart beat (from right side)

diastole

systole

pulmonary artery

vena cava

semi-lunar **valves** closed

semi-lunar valve open

right ventricle relaxed

tricuspid valves open

right ventricle contracted

tricuspid valves close

adrenalin. The rate of human heart beat is measured by counting **pulse rate**.

Herbivore An animal which feeds on plants. Herbivores include sheep, rabbits, cattle, and have a **dentition** adapted for chewing vegetation and a **gut** capable of **cellulose digestion**.

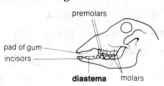

premolars

pad of gum

incisors

diastema molars

dental formula incisors ⁰⁄₄ ; canines ⁰⁄₀ premolars ; ³⁄₃ molars ³⁄₃ (total : 32)

Most herbivore **teeth** are grinders, *canines* are usually absent. In herbivores without upper *incisors*, a horny pad of gum combines with the lower incisors in biting vegetation. In most herbivores the lower jaw can move sideways, or backwards and forwards, thus causing the grinding action of the teeth. An example is a sheep, shown below.

See **Teeth**.

Heterotrophic (Holozoic) Describing organisms which obtain **organic compounds** (food) by feeding on other organisms. Heterotrophs include all animals and fungi, most **bacteria** and a few flowering plants. Heterotrophs are also called **food consumers** and can be classified into **carnivores, herbivores, omnivores, saprophytes, parasites**. See **Autotrophic**.

Heterozygous (Hybrid) Having two different **alleles** of the same **gene**. See **Monohybrid inheritance**.

Holophytic See **Autotrophic**.

Holozoic See **Heterotrophic**.

Homeostasis The maintenance of constant conditions within an organism. Examples:
Control of **blood glucose** level by **insulin**.
Control of blood water content by **A.D.H.**
Control of body temperature by the **skin**, etc.

Homiothermic Animals which maintain a constant narrow range of body temperature despite **environmental** fluctuations. *Mammals* and *birds* are homiothermic, although often described as *warm-blooded*. See **Poikilothermic; Temperature regulation**.

Homologous chromosomes Chromosomes pairs containing similar **genes**, and which come together during **meiosis**. Homologous pairs of chromosomes are found in all **diploid** organisms, one of the pair coming from the *male* **gamete** and the other from the *female* **gamete**, the pair being united at **fertilization**.

Homozygous (Pure) Having two identical **alleles** for any one **gene**. See **Monohybrid inheritance**.

Hormones (animal) Chemicals secreted by the **endocrine glands** and transported via the bloodstream to certain **organs** (*target organs*) where they cause specific effects which are vital in regulating and coordinating body activities. Hormone action is usually slower than nervous stimulation.
The table below summarizes the properties of some important human hormones; there are many others.

Endocrine gland	Hormone	Effects
Pituitary gland	A.D.H. (Anti-Diuretic Hormone)	Controls water reabsorption by the kidneys.

Endocrine gland	Hormone	Effects
	T.S.H. (Thyroid Stimulating Hormone)	Stimulates *Thyroxine* production in the thyroid gland.
	F.S.H. (Follicle-Stimulating Hormone)	Causes **ova** to mature and the **ovaries** to produce **oestrogen**.
	L.H. (Luteinizing Hormone)	Initiates **ovulation** and causes the **ovaries** to release **progesterone**.
	Growth hormone	Stimulates growth in young animals. In humans, deficiency causes *dwarfism*, and excess causes *gigantism*.
Thyroid gland	*Thyroxine*	Controls rate of growth and development in young animals. In human infants, deficiency causes *cretinism*. Controls the rate of chemical activity in adults. Excess causes thinness and over-activity, and deficiency causes obesity and sluggishness.

100

Pancreas (*Islets of Langerhans*)	Insulin	Stimulates conversion of glucose to glycogen in the liver. Deficiency causes *diabetes*.
Adrenal glands	*Adrenaline*	Under conditions of "*fight, flight, or fright*" causes changes which increase the efficiency of the animal. For example, increased heart beat and breathing, diversion of blood from gut to muscles, conversion of glycogen in the liver to glucose.
Ovaries	Oestrogen	Stimulates secondary sexual characteristics in the female, for example, breast development. Causes the uterus wall to thicken during menstrual cycle.
	Progesterone	Prepares uterus for implantation.

Endocrine gland	Hormone	Effects
Testes	*Testosterone*	Stimulates secondary sexual characteristics in the male, for example, facial hair.

Hormones (plant) Growth substances, for example, **auxins**, involved in many plant processes, including **tropisms**, **germination**, etc.

Humerus **Bone** of the upper fore-limb of *tetrapods*. In man, the bone of the upper arm. See **Endoskeleton**.

Humus Dark-coloured **organic** material in **soil** consisting of decomposing plants and animals, and providing nutrients for plants, and ultimately for animals. See **Soil**.

Hybrid See **Heterozygous**.

Hydrolysis The breakdown of complex **organic compounds** by **enzyme** action involving the addition of *water*. Hydrolysis is the basic reaction of virtually all processes of **digestion** of **proteins**, **fats**, **polysaccharides** and many other compounds.

$$\underset{\substack{\text{(for example, }\textbf{starch})}}{\overset{\text{large}}{\text{complex compound}} + H_2O} \overset{\text{enzyme}}{\longrightarrow} \underset{\substack{\text{(for example }\textbf{glucose})}}{\overset{\text{small sub-units}}{}}$$

See **Condensation**.

Hydrotropism Tropism relative to water. Plant **roots** are *positively hydrotropic*, i.e., they grow towards water.

Ileum Final region of mammalian **small intestine** which receives food from the **duodenum**. The lining of the ileum secretes **enzymes** which complete the **digestion** of **protein**, **carbohydrate** and **fat**, into **amino acids**, simple sugars (mainly **glucose**), fatty acids, and glycerol.

Absorption of food occurs in the ileum which has a large absorbing surface due to the presence of thousands of finger-like structures called **villi**. The lining of each villus is very thin, allowing the passage of soluble foods and each contains a network of **blood capillaries**.

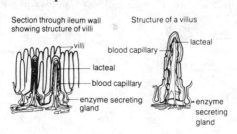

Section through ileum wall showing structure of villi

- villi
- lacteal
- blood capillary
- enzyme secreting gland

Structure of a villus

- lacteal
- blood capillary
- enzyme secreting gland

Amino acid and glucose particles diffuse into the blood capillaries from where they are transported first to the **liver** and then to the general circulation.

Fatty acid and glycerol particles pass into the **lacteals** and are circulated via the **lymphatic system**. Material not absorbed, for example, **roughage**, is passed into the **large intestine**. See **Digestion; Assimilation**.

Imago An adult, sexually mature insect. See **Metamorphosis**.

Implantation Attachment of mammalian **embryo** to the **uterus** lining at the start of **pregnancy**. In preparation for implantation, the uterus wall becomes thicker with new **cells** and an increased **blood** supply. See **Fertilization in man; Pregnancy**.

Incomplete dominance Genetic condition in which neither of a pair of **alleles** is **dominant** but

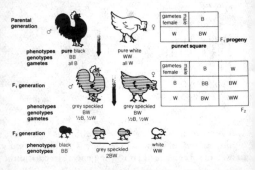

instead 'blend' to produce an intermediate trait. For example, in *Andalusian fowl* the inheritance of feather colour shows incomplete dominance.

See **Monohybrid inheritance**.

Indicator organism Organism which can survive only in certain environmental conditions, and hence one whose presence provides information about the **environment** in which it is found. For example, the **bacterium**, *Escherichia coli* lives in animal **guts** and is always present in **faeces**. Although, *E. coli* is itself harmless, its presence in water indicates sewage **pollution**.

Inorganic compounds Chemical substances within **cells** which are derived from the external physical **environment**, and which are not organic. The most abundant cell inorganic compound is *water* which is present in amounts ranging from five to ninety per cent.

The other inorganic components of cells are **mineral salts** present in amounts ranging from one to five per cent. See **Organic compounds**.

Insulin Vertebrate **hormone** secreted by the Islets of Langerhans in the **pancreas**. Insulin regulates the conversion of **glucose** to **glycogen** in the **liver**. If the concentration of **blood** glucose is high, the rate of secretion of insulin is high, and thus glucose is rapidly converted to liver glycogen. If the concentration of blood glucose is low, less insulin is secreted, an example of feedback regulation found in many hormones.

Integument 1) External protective covering of an animal, for example, **skin**, **cuticle**.

2) Protective layer around flowering plant **ovules** which after fertilization forms the **testa**.

Intercostal muscles Muscles, positioned between the ribs of mammals, important in **lung** ventilation. See **Breathing**.

Intestine Region of the **alimentary canal** between the **stomach** and the **anus** or **cloaca**. In vertebrates it is the major area of **digestion** and **absorption** of food, and is usually differentiated into an anterior **small intestine** and a posterior **large intestine**. See **Digestion**.

Involuntary (smooth) muscles Muscles associated with internal **tissues** and **organs** in mammals, for example, the **gut** and **blood vessels**, and so-called because they are not directly controlled by the will of the organism. Involuntary muscle actions include contraction and dilation of the *pupil* by the **iris** in the **eye** and **peristalsis**. See **Voluntary muscles**, **Antagonistic muscles**.

Iris Structure in the vertebrate **eye** which controls the size of the *pupil* and hence the amount of light entering the eye. See **Eye**.

Irritability See **Sensitivity**.

Joint The point in a **skeleton** where two or more **bones** meet and movement may be possible. Moveable joints in mammals are of three types:

1) *Ball and socket joints* allow movement in several planes.

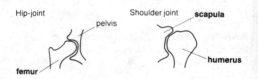

2) *Hinge joints* allow movement in only one plane.

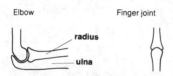

3) *Gliding joints* occur when two flat surfaces glide over one another, allowing a small amount of movement only.

joint between two vertebrate

cartilage
disc

vertebrae

Section through a moveable joint

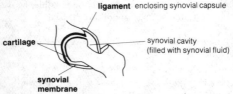

ligament enclosing synovial capsule

cartilage

synovial cavity
(filled with synovial fluid)

**synovial
membrane**

Keratin Strong, fibrous **protein** present in vertebrate **epidermis** forming the outer protective layer of **skin** and also *hair, nails, wool, feathers,* and *horns*.

Kidney Organ of **excretion** and **osmoregulation** in vertebrates, consisting of units called **nephrons**. In Man, the kidneys are a pair of red – brown oval structures at the back of the **abdomen**.

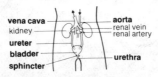

vena cava
kidney

ureter
bladder
sphincter

aorta
renal vein
renal artery

urethra

Oxygenated **blood** enters each kidney via the *renal artery*, and the *renal vein* removes *deoxygenated blood*. Another tube, the **ureter** connects each kidney with the **bladder**.

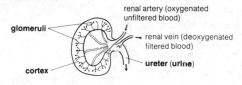

The renal artery divides into numerous *arterioles* which terminate in tiny knots of blood **capillaries** called **glomeruli**. Each glomerulus (about one million in a human kidney), is enclosed in a cup-shaped organ called a **Bowman's capsule**.

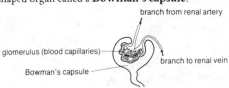

Two processes occur in the kidneys:

1) *Ultra-filtration*

2) *Reabsorption*

Ultra-filtration The vessel leaving each glomerulus is narrower than the vessel entering, causing the blood in the glomerulus to be under high pressure, which results in the blood components with smaller molecules being forced through the **selectively permeable** capillary wall into Bowman's capsule.

Large particles (unfiltered)	*Small particles* (filtered)
Blood cells	**Glucose**
Plasma proteins	**Urea**
	Mineral salts
	Water
	Amino acids

Reabsorption The fluid filtered from the blood (filtrate) passes from Bowman's capsule down the *renal tubule* where reabsorption of useful materials

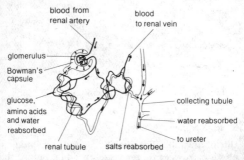

blood from renal artery

blood to renal vein

glomerulus

Bowman's capsule

glucose, amino acids and water reabsorbed

collecting tubule

water reabsorbed

to ureter

renal tubule

salts reabsorbed

occurs, i.e., *all* the glucose and amino acids, and *some* of the salts and water are re-absorbed into the blood.

The liquid resulting from filtration and reabsorption by the kidney is a solution of salts and urea in water (**urine**) and is passed to the **bladder** via the **ureters** from where it is expelled via the **urethra** under the control of a **sphincter muscle**. See **A.D.H.**

Kingdom Any of the three great divisions of living organisms, i.e., *Animal*, *Plant*, and *Protista* kingdoms. See **Classification**.

Lacteals **Lymph** vessels within the **villi** of the vertebrate **intestines**. The products of **fat digestion** (*fatty acids* and *glycerol*) diffuse into the lacteals and are circulated via the **lymphatic system**.

Lactic acid **Organic** acid (CH_3 CHOH COOH) produced during **respiration** in many animal **cells**, including vertebrate **muscle** cells, and certain **bacteria**. See **Respiration**, **Oxygen debt**.

Large intestine Posterior region of vertebrate **intestine**. In man, at the entry to the large intestine, there is a region called the **caecum**, from which arises the **appendix**, but most of the large intestine consists of the **colon** which leads to the **rectum**. The large intestine receives undigested material from the **ileum**. See **Digestion**.

Larva Intermediate, sexually immature stage in the **life history** of some animals between hatching from the egg and becoming adult, for example, amphibian *tadpoles* and butterfly *caterpillars*. See **Metamorphosis**.

Larynx Region at the upper end of the **trachea** of tetrapods opening into the **pharynx** and specialized, to close the **glottis** during swallowing. In mammals, amphibians, and reptiles, *vocal cords* within the larynx produce sound. See **Breathing in mammals**.

Leaf That part of a flowering plant, which grows from the **stem** and is typically flat and green. *Functions are*: (1) **Photosynthesis**, (2) **gas exchange**, (3) **transpiration**.

Structure (**dicotyledon**)

vertical section

upper **epidermis**

palisade mesophyll
spongy mesophyll
lower **epidermis**

xylem
phloem } **vascular bundle**

air space   **guard cell**

stoma

Lens Transparent structure in the **aqueous humour** of the vertebrate **eye**, important in focussing the image on the **retina**, and in **accommodation**. See **Eye; Accommodation**.

Lenticel One of many pores developing in woody **stems** and **roots** when **epidermis** is replaced by *bark*, through which **gas exchange** occurs. See **Gas Exchange (plants)**.

Leucocyte See **White blood cell**.

L.H. (luteinizing hormone) **Hormone** secreted by the vertebrate **pituitary gland**. See **Hormones, Ovulation**.

Lichen Plant formed by a **mutualistic** relationship between an *alga* and a *fungus*. The alga supplies **carbohydrate** and oxygen to the fungus, and receives water and **mineral salts** in return.

Life history (life cycle) The various stages of development which organisms undergo from egg to adult. See **Alternation of generations; Metamorphosis**.

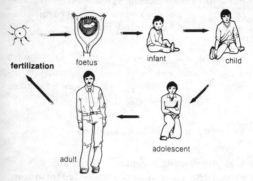

fertilization foetus infant child

adult adolescent

Ligament Strong band of **collagen** connecting the **bones** at moveable vertebrate **joints**. Ligaments strengthen the joint, allowing movement in only certain directions and preventing dislocation. See **Joint**.

Lignin **Organic compound** deposited in the **cell** walls of **xylem** vessels, giving strength. Lignin is an important constituent of *wood*. See **Xylem**.

Limiting factor Any factor of the **environment** whose level at a particular time inhibits some activity of an organism or **population** of organisms. For example, consider the effects of *temperature* and *light intensity* on the rate of **photosynthesis**.

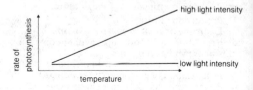

Increasing temperature has little effect at low light intensities. Thus, in this case, light intensity must be the limiting factor.

Lipase Enzyme which digests **fat** into *fatty acids and glycerol* by **hydrolysis**. In mammals, lipase is secreted by the **pancreas** and the **ileum**.

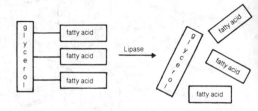

Lipid See **Fat**.

Liver Largest organ of the vertebrate body, occupying much of the upper part of the **abdomen**, in close association with the **alimentary canal**. See **Digestion; Circulatory systems**.

Some of the many functions of the liver are:

1) Production of **bile**.
2) **Deamination** of excess **amino acids**.
3) Regulation of **blood** sugar by interconversion of **glucose** and **glycogen**.
4) Storage of iron, and **vitamins** A and D.
5) Detoxication of poisonous by-products.
6) Release and distribution of heat produced by the chemical activity of liver cells.
7) Conversion of stored **fat** for use by the **tissues**.
8) Manufacture of **fibrinogen**.

Long sight (Hypermetrophia) Human **eye** defect mainly caused by the distance from **lens** to

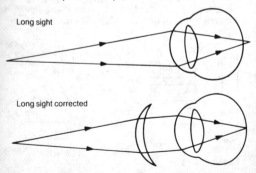

Long sight

Long sight corrected

retina being shorter than normal. This results in *near* objects being focussed behind the retina giving blurred vision. Long sight is corrected by wearing *converging (convex)* lenses.

Lungs **Breathing organs** of mammals, amphibians, reptiles, and birds. In mammals, the lungs are two elastic sacs in the **thorax** which can be expanded or compressed by movements of the thorax in such a way that air is continually taken in and expelled. The **trachea** (windpipe) connects the lungs with the atmosphere. It divides into two **bronchi** which enter the lungs and further divide into many smaller *bronchioles* which terminate into millions of air-sacs called **alveoli** which are the **gas exchange** surface and which are in close contact with **blood vessels** bringing blood from and to the **heart**.

Air passages in the lungs

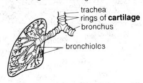

trachea
rings of **cartilage**
bronchus

bronchioles

See **Breathing in mammals; Gas exchange (mammals)**.

Lymph Fluid drained from **blood capillaries** in vertebrates as a result of high pressure at the arterial end of the capillary bed. Lymph or *tissue fluid* which is similar to **plasma** (except for a much lower concentration of **plasma proteins**) bathes the **tissues** and acts as a medium in which substances are exchanged between capillaries and **cells**. For example, *oxygen* and *glucose* diffuse into the cells while *carbon dioxide* and **urea** are removed. Lymph drains back into capillaries or into vessels called *lymphatics* which then connect with the general circulation via the **lymphatic system**.

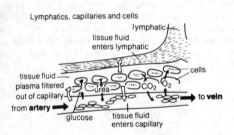

Lymphatics, capillaries and cells

Lymphatic system   System of fluid-containing vessels (*lymphatics*) in vertebrates, which return **lymph** to the general **blood** circulation. The lymphatic system is also important in:

1) Transporting the products of **fat digestion**.

2) Production of **white blood cells** and **antibodies**.

Lymph nodes Structures within the **lymphatic system** which filter **bacteria** from **lymph** and produce **white blood cells** and **antibodies**.

Medulla The central part of a **tissue** or **organ**, for example, mammalian **kidney**. See **Cortex**.

Medulla oblongata (medulla) Posterior region of vertebrate **brain** which is continuous with the **spinal cord** and which in mammals, controls **heart beat**, **breathing**, **peristalsis** and other *involuntary actions*. See **Brain**.

Meiosis (reduction division) Two successive **cell divisions** which produce **gametes** containing *half* the normal **chromosome** number (the **haploid** number).

Meiosis is necessary in all sexually reproducing organisms, so that at **fertilization,** the normal (**diploid**) chromosome number is restored.

Meiosis produces **spermatazoa** and **ova** in animals and **pollen** grains and **embryo sacs** in flowering plants and is particularly important because it allows new combinations of **genes** so allowing heritable **variation** in a **species**.

The stages in meiosis

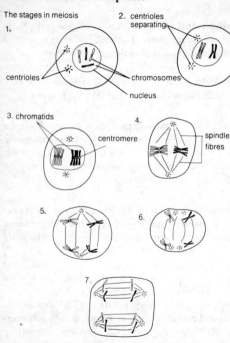

1.

centrioles

chromosomes

nucleus

2. centrioles separating

3. chromatids

centromere

4.

spindle fibres

5.

6.

7.

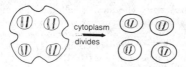

Menstrual cycle Reproductive cycle occurring in female *primates* (monkeys, apes, humans). In human females, the cycle lasts about twenty-eight days, during which the **uterus** is prepared for **implantation**. If **fertilization** does not occur, the new uterus lining and unfertilized **ovum** are expelled, which results in bleeding from the **vagina** (*menstruation*). See **Fertilization in man; Ovulation**.

Meristem Localized **tissue** of active **cell division** which is responsible for **growth** in plants. The **cells** at meristems are *undifferentiated*, but by repeated cell divisions, new cells are produced which ultimately *differentiate* to form the specialized tissues of plants, for example, **xylem**, **phloem**, etc. Meristematic activity is controlled by plant **hormones** and the principal meristems are **root** tip, **shoot** tip and **cambium**. See **Cell differentiation; Primary growth; Secondary growth**.

Metabolic water One of the products of aerobic **respiration** which is an important source of water for desert animals.

Metabolism The sum of all the physical and chemical processes occurring within a living organism. These include both the synthesis (*anabolism*) and breakdown (*catabolism*) of compounds and is measured by *basal metabolic rate* (**B.M.R.**).

Metamorphosis The period in the **life history** of some animals when the *juvenile* stage is transformed into an adult. For example, amphibians:

Incomplete metamorphosis A type of development in which there are relatively few changes from juvenile form to adult. It occurs in insects such as *dragonfly*, *locust* and *cockroach*, in which the juvenile form (**nymph**) resembles the adult except that it is smaller, wingless, and sexually immature. For example, cockroach:

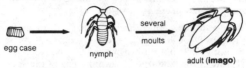

egg case

nymph

several moults

adult (**imago**)

Complete metamorphosis Involves great changes from **larva** to adult. It occurs in insects such as *butterfly*, *moth*, *housefly*, etc., and the **larvae** in such life histories are *maggots*, *grubs*, or *caterpillars* (depending on the species) and are quite unalike the adult form. A series of moults (**ecdyses**) produces the **pupa** which then becomes completely reorganized and develops into the adult, the only sexually mature stage. For example, butterfly:

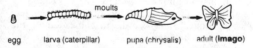

egg

larva (caterpillar)

moults

pupa (chrysalis)

adult (**imago**)

Micro-organisms Very small living organisms which can usually only be seen with the aid of a **microscope**. Micro-organisms include *protozoans* *algae*, **viruses**, *fungi*, **bacteria**.

Micropyle a) Pore in a **seed** through which water is absorbed at the start of **germination**.

b) Pore in the **ovule** of a **flower** through which the **pollen** tube delivers the male **gamete**.

c) Pore in the **ovum** of insects through which the **spermatozoon** enters.

Microscope An instrument used to magnify structures, for example, **cells** or organisms, which are not visible to the naked eye.

The light microscope Light illuminates the specimen which is magnified by glass lenses.

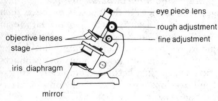

The magnification of the microscope is found by multiplying the magnification of the *objective lens* (for example, × 40) by the magnification of the *eyepiece lens* (for example, × 10) to give the total magnification (in this example × 400). The maximum possible magnification using a light microscope is × 1500. Thin specimens are placed on a glass slide and may be *stained* with dyes which show up particular structures.

The phase contrast microscope allows the viewing of transparent and unstained structures.

The electron microscope is the most advanced type, giving magnification as high as × 500 000.

Milk teeth (deciduous teeth) The first of two sets of **teeth** occurring in most mammals. For example, humans have *twenty* milk teeth which are replaced during childhood by the larger *permanent* teeth.

Mineral salts Components of **soil** formed from rock *weathering* and **humus** *mineralization* and found in **solution** in soil water. Mineral salts are absorbed by plant **roots** and transported through the plant in the **transpiration stream**. Like **vitamins**, mineral salts are required in tiny amounts, but are nevertheless vital for plant and ultimately animal nutrition; the absence of a particular mineral salt can lead to *mineral deficiency disease* and death. Plants require at least twelve mineral salts for healthy growth.

Essential elements required in relatively large quantities: *nitrogen, phosphorus, sulphur, potassium, calcium, magnesium.*

Trace elements required in very small amounts: *manganese, copper, zinc, iron, boron, molybdenum.*

Some mineral salts are required by plants some are required by animals, and some by both. The

125

properties of some important mineral salts are summarized in the table below.

Mineral salt	Function	Some effects of deficiency
Phosphorus	Component of A.T.P., nucleic acids, *cell membrane*; animal bones.	Stunted plant growth.
Calcium	Component of plant *cell walls* and animal bones.	*Rickets* in humans
Nitrogen	Component of **protein** and nucleic acids.	Poor reproductive development in plants.
Iron	Component of **haemoglobin**.	*Anaemia* in humans.
Magnesium	Component of *chlorophyll*.	Pale yellow plant leaves (*chlorosis*).

Mitochondrion Microscopic **cell** organelle in the **cytoplasm** of **aerobic** cells, in which some **respiration** reactions occur. The inner membrane of a mitochondrion wall is highly folded giving rise to a series of partitions called *cristae* which greatly increases the surface area for the attachment of respiratory **enzymes**.

Mitosis The process by which the **nucleus** of a **cell** divides in such a way that the resultant *daughter* cells receive precisely the same numbers and types of **chromosomes** as the original *mother cell*. During this type of **cell division** the chromosomes of the mother cell (the dividing cell) are first duplicated and then passed in identical sets to the two daughter cells.

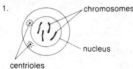

Chromosomes appear as coiled threads. Two structures called *centrioles* are seen outside the nucleus.

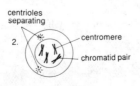

Each chromosome makes a duplicate of itself. These duplicates or *chromatids* are joined at a *centromere*.

3.

The *nuclear membrane* disappears. The centrioles move to opposite poles of the cell and produce a system of fibres called a *spindle*. The chromatid pairs line up at the equator of the cell.

4.

The chromatid pairs separate and move to opposite poles of the cell.

5.

New nuclear membranes form and the **cytoplasm** starts to divide.

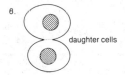

The chromosomes disappear and the cells return to the resting state.

Monocotyledons Smaller of the two sub-sets of flowering plants, the other being **dicotyledons**. The characteristics of monocotyledons are:

One **cotyledon** in the **seed**.
Parallel **veins** in **leaves**.
Narrow leaves.
Scattered **vascular bundles** in **stem**.
Examples: cereals, grasses.

Monohybrid inheritance The inheritance of one pair of contrasting characteristics.

For example, in the fruit fly *Drosophila*, one variety is **pure-breeding** *normal winged*, and another is pure-breeding *vestigial winged*, the normal winged **allele** being **dominant** to the vestigial winged allele. The diagrams below show the result of crossing normal winged females with vestigial winged males.

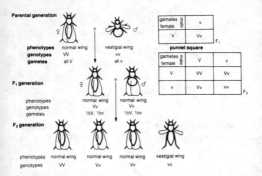

	Parental generation		
phenotypes	normal wing	vestigial wing	
genotypes	VV	vv	
gametes	all V	all v	

punnet square

gametes female \ male	v
V	Vv

F_1

gametes female \ male	V	v
V	VV	Vv
v	Vv	vv

F_2

	F₁ generation	
phenotypes	normal wing	normal wing
genotypes	Vv	Vv
gametes	½V, ½v	½V, ½v

	F₂ generation			
phenotypes	normal wing	normal wing	normal wing	vestigial wing
genotypes	VV	Vv	Vv	vv

Approximate ratio in F_2 generation:

$$\frac{\text{normal wing}}{\text{vestigial wing}} = \frac{3}{1}$$

Explanation When normal winged is crossed with vestigial winged, the vestigial trait seems to disappear, only to reappear again to a limited extent in the next generation, suggesting that the **F_1 generation** must have possessed this trait without showing it. A trait such as normal wing which always appears in a cross between contrasting parents is described as **dominant**, while a trait such as vestigial wing which is 'lost' in F$_1$ generation **progeny**, apparently masked by a dominant trait is called **recessive**.

For each trait, an organism receives one **gene** from the male **gamete** and one from the female gamete, i.e., the **zygote** and resulting organism contain two genes for every trait, but the gametes contain only one, (as a result of **meiosis**). If the paired genes for a particular trait are identical, the organism is said to be **homozygous** or **pure** for that trait. When an organism has two different genes for a trait, it is described as **heterozygous** or **hybrid**. Alternative forms of a gene are described as *allelomorphs* or **alleles**. Hence, if the allele for normal wing is represented by V and that for vestigial wing by v:

VV is homozygous normal wing
Vv is heterozygous normal wing
vv is homozygous vestigial wing

Organisms, homozygous for a particular trait, produce only one type of gamete for that trait, while heterozygous organisms produce two gamete types. The results of **fertilization** can then be worked out using a **punnet square**. See **Backcross; Incomplete dominance**.

Monosaccharides Single sugar **carbohydrates** which are the sub-units of more complex carbohydrates, and named on the basis of the number of *carbon* atoms present. For example:

$C_3H_6O_3$	$C_5H_{10}O_5$	$C_6H_{12}O_6$
triose sugar	pentose sugar	hexose sugar

Hexose sugars are common carbohydrates and include **glucose**.

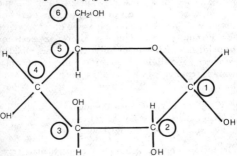

structure of glucose ($C_6H_{12}O_6$)

Mouth The opening in an animal's head through which food is taken (*ingested*) and through which sounds are uttered. In man, food is broken down into smaller pieces by the action of the **teeth** and is mixed with **saliva**, a fluid secreted by the *salivary glands*.

Mucus Slimy fluid secreted by **goblet cells** in vertebrate **epithelia**. Mucus traps dust and **bacteria** in mammalian air passages, lubricates the surfaces of internal organs, and facilitates the movement of food through the **gut** while preventing the digestive enzymes from reaching and digesting the gut itself.

Multicellular (organism) An organism consisting of *many* **cells**. This includes most animals and plants, for example, an adult man consists of 10^{14} cells. See **Unicellular**.

Muscle Animal **tissue** consisting of **cells** which are capable of *contraction* as a result of **nerve impulses**, thus producing movement, both of the organism as a whole and of internal **organs**.

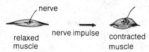

relaxed muscle   nerve impulse   contracted muscle

See **Antagonistic muscles**; **Involuntary muscles**; **Voluntary muscles**.

Mutation Spontaneous change in the structure of **D.N.A.** in **chromosomes**. Mutations occur rarely but when they do, they are inheritable and most confer disadvantages on the organisms inheriting them. Mutations can result in beneficial **variations** within a **population** which can lead to **evolution**, and although they occur naturally, they can also be induced by exposure to excessive *radiation*.

Mutualism   A **symbiotic** relationship in which both organisms benefit. For example, the **intestinal bacteria** of **herbivores** digest the **cellulose** of plant **cell** walls, the products of which are then used by the herbivore.

Natural selection   The mechanism proposed by *Charles Darwin* to suggest how **evolution** could have taken place. Darwin suggested that individuals in a **species** differ in the extent to which they are adapted to their **environment**. Thus, in **competition** for food, etc., the better adapted organisms will survive, and pass on their favourable **variations**, while the less well adapted will be eliminated.

Nephron   Sub-unit of the vertebrate **kidney**, consisting of a **Bowman's capsule**, **glomerulus**, and *renal tubule*. See **Kidney**.

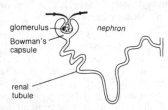

glomerulus

nephron

Bowman's capsule

renal tubule

Nerve cells (neurones) Cells which are the basic units of mammalian **nervous systems**. There are two types of nerve cells:

Sensory neurones conduct **nerve impulses** from **receptors** to the **central nervous system** (C.N.S.), i.e., from **eyes**, **ears**, **skin**, etc.

Motor neurones conduct **nerve impulses** from the C.N.S. to **effectors**, such as **muscles** and **endocrine glands**.

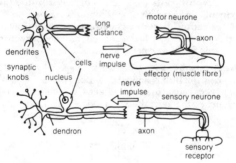

Each nerve cell consists of three parts:

1) A *cell-body* containing **cytoplasm** and **nucleus** and forming the *grey matter* in the **brain** and **spinal cord**.

2) Fibres which carry nerve impulses *into* cell-bodies. In sensory neurones, this fibre is a single *dendron* while in motor neurones there are numerous *dendrites*. In the C.N.S. such fibres form *white matter*.

3) Fibres called **axons** which carry nerve impulses *from* cell-bodies.

Nerve impulses The electrical messages by which information is transmitted rapidly throughout **nervous systems**. Nerve impulses are initiated at **receptor cells** as a result of **environmental stimuli**. In vertebrates, the impulses are conducted to the **central nervous system**, where they trigger other impulses which are relayed to **effector organs**. See **Nerve cells**, **Synapse**.

Nervous system Network of specialized **cells** in **multicellular** animals, which acts as a link between **receptors** and **effectors**, and thus co-ordinates the animal's activities. In mammals, the

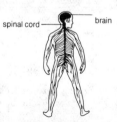

spinal cord — brain

nervous system consists of the **brain** and **spinal cord** (which together form the **central nervous system**) and **nerve cells** connecting to all parts of the body.

Neurone   See **Nerve cell**.

Niche   The status or way of life of an organism within a **community**. For example, a **herbivore** and a **carnivore** may share the same **habitat** but their different feeding methods mean that they occupy different *niches*

Nitrification   The conversion by **soil** *nitrifying bacteria* of **organic** *nitrogen* compounds, for example, *ammonia*, into *nitrates* which can be absorbed by plants. Ammonia is first converted to *nitrites* by *Nitrosomonas* **bacteria** and the nitrites to nitrates by *Nitrobacter* **species**. See **Nitrogen cycle**.

Nitrogen cycle   The circulation of the element *nitrogen* and it compounds in nature, caused mainly by the **metabolic** processes of living organisms. The nitrogen cycle is summarized below.

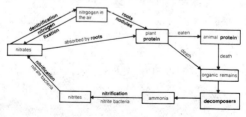

Nitrogen fixation The conversion of atmospheric *nitrogen* by certain **micro-organisms** into **organic** nitrogen compounds. Nitrogen-fixing **bacteria** live either in **soil** air, or within the **root nodules**, of *leguminous plants*. The activity of these organisms, for example, *Azotobacter, Rhizobium*, enriches the soil with nitrogen compounds. See **Nitrogen cycle; Root nodule**.

Normal distribution curve Bell-shaped curve obtained when *continuous variation* is measured in a **population**. See **Variation**.

Nucleic acids **Organic compounds** found in all living organisms, particularly associated with the **nucleus** of the **cell**, and consisting of sub-units called *nucleotides*.

Nucleotide

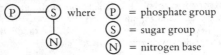 where P = phosphate group
S = sugar group
N = nitrogen base

The sugar group of one nucleotide can combine with the phosphate group of another to form a *polynucleotide* chain:

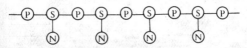

Such polynucleotide chains are the basis of nucleic acid structure. See **D.N.A.**, **R.N.A.**

Nucleus Structure within most **cells** in which the **chromosomes** are located. It is isolated from the **cytoplasm** by a *nuclear membrane*, and chromosomes are visible only during **cell division**. As the chromosomes contain the hereditary information, the nucleus controls all the cell's activities through the action of the genetic material **D.N.A**

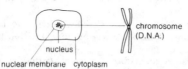

chromosome (D.N.A.)

nucleus

nuclear membrane   cytoplasm

Nymph Juvenile form of certain insects which resembles the **imago** except that it is smaller, wingless and sexually immature. See **Metamorphosis**.

Oesophagus Region of the **alimentary canal**, connecting the mouth with the digestive areas. In vertebrates, it runs between the **pharynx** and the **stomach** and transports food by **peristalsis**. See **Digestion**.

Oestrogen **Hormone** secreted by vertebrate **ovaries** which stimulates the development of **secondary sexual characteristics** in female mammals and is important in the **menstrual cycle**.

Omnivore   An animal which feeds on both plants and animals. Omnivores include man whose **dentition**, like other omnivores, contains both **herbivore** and **carnivore** features, consisting of biting, ripping and grinding teeth, which suit the mixed diet. See **Dental formula**; **Teeth**.

Optic nerve   Cranial nerve of vertebrates conducting **nerve impulses** from the **retina** to the **brain**. See **Eye**.

Order   Unit used in the **classification** of living organisms consisting of one or more **families**.

Organ   A collection of different **tissues** in a plant or animal which form a structural and functional unit, for example, the **liver**, a plant **leaf**.

Different organs may then be associated together to constitute a *system*, for example, the digestive system.

cells → tissues → organs → systems

Organ of corti   See **Cochlea**.

Organic compounds   Compounds containing the element *carbon*, found in all living organisms. The major organic compounds are **carbohydrates**, **fats**, **nucleic acids**, **proteins**, and **vitamins**. see **Inorganic compounds**.

Osmoregulation The method by which animals maintain the *water* and **mineral salt** content of their body fluids at the correct level.

Examples of osmoregulation

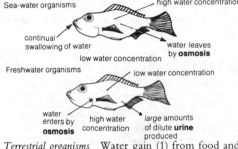

Sea-water organisms

high water concentration

continual swallowing of water

water leaves by **osmosis**

low water concentration

Freshwater organisms

low water concentration

water enters by **osmosis**

high water concentration

large amounts of dilute **urine** produced

Terrestrial organisms Water gain (1) from food and drink (2) as a by-product of **respiration**.

Water loss (1) by sweating (2) in exhaled air (3) as urine.

Water and mineral salt balance in terrestrial animals is mainly under the control of the **kidneys**.

Osmosis The **diffusion** of *solvent* (usually water) particles through a **selectively permeable**

membrane from a region of high solvent concentration to a region of lower solvent concentration. For example:

	Selectively permeable membrane	
pure water	diffusion →	20% sugar solution
(100% H_2O)		(80% H_2O)
10% sugar solution	diffusion →	20% sugar solution
(90% H_2O)		(80% H_2O)
pure water	← equilibrium →	pure water

Examples of selectively permeable membranes are: (1) The **cell** membrane (2) visking (dialysis) tubing. Such membranes are thought to have tiny pores which allow the rapid passage of small water particles, but restrict the passage of larger *solute* particles.

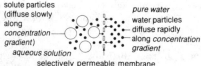

solute particles (diffuse slowly along *concentration gradient*)

aqueous solution

pure water
water particles diffuse rapidly along *concentration gradient*

selectively permeable membrane

Since the *cell membrane* is selectively permeable, osmosis is important in the passage of water into and out of cells and organisms.

Osmotic pressure   Osmosis can be demonstrated using an *osmometer*.

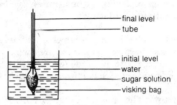

Water moves into the visking bag by osmosis, causing the liquid level in the tube to rise. The pressure exerted by the osmotic movement of water is called *osmotic pressure* and depends on the relative *solute* concentrations of the **solutions** involved. The osmotic pressure that a solution is capable of developing is called its *osmotic potential*, but is only realized in an osmometer.

Ossicles   Chain of three tiny **bones** in the mammalian *middle ear*. See **Ear**.

Oval window   Membrane separating the *middle* and *inner ear* in mammals. See **Ear**.

Ovary 1) Hollow region in the **carpel** of a **flower**, containing one or more **ovules**. See **Fertilization in plants**.

2) Reproductive **organ** of female animals. In vertebrates, there are two ovaries which produce the **ova** and also release certain sex **hormones**. See **Fertilization in man**; **Ovulation**.

Oviduct Tube in animals which carries **ova** from the **ovaries**. In mammals there are two oviducts leading to the **uterus** and **fertilization** occurs within the **oviduct**. See **Fertilization in man**.

Ovulation The release of an **ovum** from a mature **graafian follicle** on to the surface of a vertebrate **ovary**, from where it passes into the **oviduct** and then into the **uterus**.

Ovulation in the human female Ovulation is controlled by **hormones** from the **pituitary gland** and the sequence of events in the female's reproductive behaviour is called the **menstrual cycle**. **Follicle stimulating hormone** (F.S.H.) induces the maturation of **ova** and causes the ovaries to produce **oestrogen**. **Luteinizing hormone** (L.H.) triggers ovulation and also the release of **progesterone** by the ovaries.

If the mature ovum is not fertilized, it is expelled with the new uterus lining and some **blood** via the

vagina, a process called *menstruation*. The main features of the human menstrual cycle are shown below:

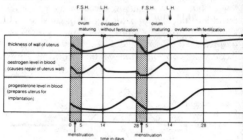

See **Fertilization in man**.

Ovule Structure in flowering plants which develops into a **seed** after **fertilization**. See **Carpel; Fertilization in plants**.

Ovum An unfertilized female **gamete** produced at the **ovary** of many animals, and containing a **haploid nucleus**. See **Fertilization; Meiosis; Ovulation**.

Oxygen debt Deficit of *oxygen* which occurs in **aerobes** when work is done with inadequate oxygen supply. For example, in mammalian **muscle** during exercise, the oxygen supply may be insufficient to meet the **energy** demand. When this happens, the

cells produce energy by **anaerobic respiration**, **lactic acid**, being a by-product.

The accumulation of lactic acid causes *muscle fatigue* but is eventually reduced as oxygen intake returns to normal after the period of exercise. This shortfall of oxygen must be repaid by increased oxygen intake (panting). The effect of exercise on lactic acid concentration of the **blood** is shown below.

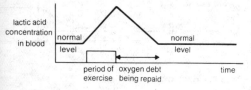

See **Respiration**.

Palisade mesophyll Main **photosynthesizing tissue** of a **leaf**, situated below the upper **epidermis**, and containing many chloroplasts. See **Leaf**.

Pancreas Gland situated near the **duodenum** of vertebrates. It releases an *alkaline* fluid into the duodenum, containing digestive **enzymes**, for example, **lipase**, **amylase**, **trypsin**. See **Digestion**.

The pancreas also contains **tissue** known as the *Islets of Langerhans*, which secretes the **hormone insulin**.

Parasite An organism that feeds in or on another living organism which is called the *host*, and which does not benefit and may be harmed by the relationship. Parasites of man include *fleas, lice, tapeworms*.

Parental generation The first organisms crossed in a breeding experiment, producing **progeny** known as the F_1 **generation**. See **Monohybrid inheritance**.

Patella Bone over the front of the knee **joint** in many vertebrates. In man it is the *knee-cap*. See **Endoskeleton**.

Pathogen A term used to describe a **parasitic** organism causing disease in another **species**.

Penis Organ in mammals by which the male gametes (**sperms**) are introduced into the female body. It also contains the **urethra** through which **urine** is discharged. See **Fertilization in man**.

Pepsin **Protease enzyme** secreted by the wall of the vertebrate **stomach**, along with *hydrochloric acid*. The acid provides a suitable *pH* for pepsin which digests long protein chains into shorter chains of **amino acids** called **peptides**.

147

Peptide Compound consisting of two or more **amino acids** linked between the *amino* group of one and the *acid* group of the next. The link between adjacent amino acids is called a *peptide bond*, and when many amino acids are joined in this way, the whole complex is called a *polypeptide*, which is the basis of **proetin** structure.

$$\underset{\text{amino acid}}{\begin{array}{c} H \quad R \quad [OH \quad H] \\ | \quad | \quad | \\ N-C-C \\ | \quad | \quad \| \\ H \quad H \quad O \end{array}} + \underset{\text{amino acid}}{\begin{array}{c} R \quad OH \\ | \quad | \\ N-C-H \\ | \quad | \\ H \quad H \quad O \end{array}} \xrightarrow{-H_2O} \underset{\text{peptide}}{\begin{array}{c} \text{peptide bond} \\ H \quad R \quad \quad R \quad OH \\ | \quad | \quad \overbrace{\quad\quad} \quad | \quad | \\ N-C-C-N-C-C \\ | \quad | \quad \| \quad | \quad | \quad \| \\ H \quad H \quad O \quad H \quad H \quad O \end{array}}$$

Peristalsis Waves of muscular contraction passing along and causing movement of contents in tubular **organs** for example, in mammals, the **alimentary canal**, and also **ureters** and **oviducts**. Peristalsis is caused by the rhythmic and co-ordinated contraction and relaxation of **involuntary** circular and longitudinal **muscles**. See **Antagonistic muscles**.

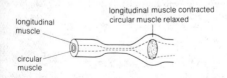

longitudinal muscle contracted
circular muscle relaxed

longitudinal muscle

circular muscle

Pest Any living organism which is considered to have a detrimental effect on man. Some examples are shown in the table below:

Effect on man	Example of pest
Reduces the growth of his plants and crops.	weeds; locusts
Causes disease in his animals	foot and mouth virus
Damages his structures	woodworm; wet rot fungus
Transmit human disease.	mosquito; lice

Methods used to combat pests include:

1) Spraying with chemicals (**pesticides**).
2) Using natural **predators** against the pest.
3) Introduce **parasites** and **pathogens** to the pest **population**.
4) Introduce *sterile* individuals to the pest population, thus reducing reproductive capacity.

Pesticide Chemical compound, often delivered in a spray, which kills or inhibits the growth of **pests**. Examples:

Herbicides Weed-killers such as *paraquat*.
Fungicides Seed-dressings such as *organo-mercury* compounds.

Insecticides Fly-sprays used in the home; sprays released from aircraft against locusts; *D.D.T.* (now banned in Britain) has been used against *mosquitos* and *lice*.

Disadvantages of pesticides

1) They may kill organisms, other than the *target test*.
2) The concentration of a pesticide increases as it passes through a **food chain**.
3) Some decompose slowly and may accumulate into harmful doses within organisms.
4) By killing off susceptible organisms, they allow resistant individuals to grow and multiply with reduced **competition**.

Phagocytosis The process by which **cells** (*phagocytes*) surround, and engulf a food particle which is then digested. Phagocytosis is the feeding method employed by some **unicellular** *protozoans*, for example, *Amoeba*. It is also one of the methods by which **white blood cells** destroy invading **micro-organisms**.

150

Pharynx Region of the vertebrate **alimentary canal** between the **mouth** and the **oesophagus**. In man it is the back of the nose and throat, and when stimulated by food, swallowing is initiated.

Phenotype The physical characteristics of an organism resulting from the influence of **genotype** and **environment**. See **Monohybrid inheritance**.

Phloem **Tissue** within plants which transports **carbohydrate** from the **leaves** throughout the plant. Phloem consists of tubes which are formed from columns of living **cells** in which the horizontal cross walls have become perforated. This allows carbohydrate to move from one phloem cell into the next and thus through the plant. Because of their structure phloem tubes are also called *sieve tubes*.

phloem cells   phloem sieve tubes

perforation of
→
cross walls

See **Leaf, Root, Secondary growth, Stem**.

Photosynthesis The process by which *green plants* make **carbohydrate** from *carbon dioxide* and *water*. The **energy** for the reaction comes from *sunlight* which is absorbed by the *chlorophyll* within

chloroplasts and *oxygen* is evolved as a by-product.

Overall reaction:

$$
\begin{array}{ccc}
\text{Carbon} & + & \text{water} \\
\text{dioxide} & & 12H_2O \\
6CO_2 & &
\end{array}
\quad
\xrightarrow[\text{chlorophyll}]{\text{light energy}}
$$

$$
\begin{array}{cccc}
\text{carbohydrate} & + & \text{water} & + & \text{Oxygen} \\
C_6H_{12}O_6 & & 6H_2O & & 6O_2
\end{array}
$$

Photosynthesis is in fact a two-stage reaction involving:

1) *The light reaction* in which light energy is used to split water into *hydrogen* (which passes to the next stage) and *oxygen* (which is released).

2) *The dark reaction* in which the hydrogen from the light reaction combines with carbon dioxide to form carbohydrates.

Summary

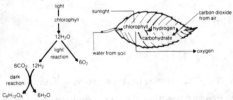

Photosynthesis is the source of all food and the basis of **food chains**, while the release of oxygen replenishes the oxygen content of the atmosphere.

Phototropism   Tropism relative to light. Plant **shoots** are *positively phototropic*, i.e., they grow towards light.

Phylum   Unit used in the **classification** of living organisms, consisting of one or more **classes**. The term *division* is often substituted in plant classification.

Pinna   Flap of **skin** and **cartilage** at the outside end of the mammalian *outer ear*. See **Ear**.

Pituitary gland   Endocrine gland at the base of the vertebrate **brain**. It produces numerous **hormones** including **A.D.H.** and **F.S.H.**, many of which regulate the activity of other endocrine glands. The pituitary gland's own secretion is in many cases regulated by the brain. See **Hormones**.

Placenta   Organ developing during **pregnancy** in the mammalian **uterus** and forming a close association between maternal and foetal **blood** circulations. The placenta allows passage of *food* and *oxygen* to the **foetus** and removes *carbon dioxide* and *urea*. See **Pregnancy**.

Plankton   Microscopic animals (*zooplankton*) and plants (*phytoplankton*) which float in the surface waters of lakes and seas. Plankton are important as the basis of aquatic **food chains**.

Plasma The clear fluid of vertebrate **blood** in which the blood **cells** are suspended. It is an aqueous **solution** in which are dissolved many compounds in transit around the body. Examples:

carbon dioxide
urea $\Big\}$ waste products;

glucose
amino acids $\Big\}$ digested foods;

hormones
plasma proteins
sodium chloride

Plasma proteins Proteins dissolved in the **plasma** of vertebrate **blood**. For example, **antibodies**, **fibrinogen**, and some **hormones**.

Plasmolysis Loss of water from a plant **cell** when the cell is surrounded by a **solution** whose water concentration is less than that of the cell **vacuole** (for example, a strong sugar or salt solution). **Osmosis** causes water to pass out of the cell, making the vacuole shrink, resulting in the **cytoplasm** being pulled away from the cell wall.

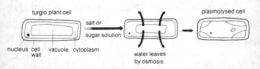

turgid plant cell   salt or sugar solution   plasmolysed cell

nucleus   cell wall   vacuole   cytoplasm   water leaves by osmosis

Plasmolysis can be induced and reversed experimentally, but continued plasmolysis results in cell death. This osmotic death rarely occurs naturally but can result from adding excess **fertilizer** to plants since this induces plasmolysis, or what is called '*plant burning*'.

Platelets Smallest **cell** of mammalian **blood**, involved in **blood clotting**.

Pleural membranes Double lining covering outside of the **lungs** and the inside of the **thorax** in mammals, and secreting *pleural fluid* between them, so facilitating **breathing** movements. See **Breathing**.

Plumule The leafy part of the embryonic **shoot** of **seed** plants. See **Seed; Germination**.

Poikilothermic Animals, whose body temperatures vary with **environmental** temperature. All animals, excluding birds and mammals, are poikilothermic, although often described as '*cold-blooded*'. See **Homiothermic**.

Pollen Reproductive spores of flowering plants, each containing a male **gamete**. Pollen grains are adapted to their mode of transfer, either by *insects* or by *wind*.

Insect pollination
spiky and sticky

Wind pollination
smooth and light

air bladders

See **Fertilization in plants; Pollination; Stamen**.

Pollination   The transfer of **pollen** grains from **stamens** to **carpels** in flowering plants. Pollination within the same **flower** or between flowers on the same plant is called *self-pollination*. Pollination between two separate plants is called *cross-pollination*. Normally male and female parts of the same plant do not mature simultaneously, favouring cross-pollination with a consequent mixing of **chromosomes** which can lead to **variation**. Pollen is transferred on the bodies of *insects* or by the *wind*.

Insect pollination   Insects visit flowers to drink or collect *nectar*. Their bodies become dusted with pollen some of which may adhere to the stigmas of subsequent flowers which they visit.

Wind pollination   Pollen grains carried by the wind must be produced in much higher numbers to compensate for loss during transfer. See **Fertilization in plants; Flowers; Pollen**.

Pollution The addition of any substance to the **environment** which upsets the natural balance. Pollution has resulted mainly from *industrialization* which is largely based on *fossil fuel* burning and which caused *migration* from the land to towns and cities.

Air pollution is caused particularly by fossil fuel burning:

coal	burning	smoke +	carbon dioxide	+ sulphur dioxide
	$\longrightarrow$			

petrol	burning	smoke +	carbon monoxide	+ oxides of nitrogen
	$\longrightarrow$			+ lead

Air pollutants such as *smoke* and *sulphur dioxide* cause irritation in the human respiratory system and may accelerate diseases such as *bronchitis* and *lung cancer*.

Water pollution results from the intentional or accidental addition of materials into both freshwater and seawater. The pollutants originate from industrial and agricultural practises and also from the home. For example, mine and quarry washings; acids; **pesticides**; oil; radioactive wastes; **fertilizers**; detergents; sewage; hot water (from power stations).

Some pollutants, such as pesticides, may poison aquatic organisms while **organic** pollutants, such as sewage, cause an increase in the **micro-organism** population in the water with a resulting decrease in *dissolved oxygen* levels, making the water unfit for many organisms. See graph below.

dissolved oxygen

bacteria

sewage
discharge

distance
downstream

Polysaccharides Carbohydrates consisting of long chains of **monosaccharides** linked together by **condensation** bonds. For example, **glucose** units can be linked in different ways to form several polysaccharides, such as **starch**, **glycogen**, **cellulose**.

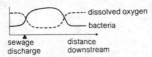

Population A group of organisms of the same **species** within a **community**.

Predator An animal that feeds on other animals which are called the *prey*, i.e., a predator is a **food consumer** (but is not a **parasite**). The relationship between predator and prey can have dramatic affects on their numbers. A typical predator – prey relationship is shown below.

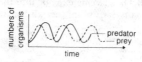

Pregnancy (Gestation period) In mammals the time from conception to **birth**. Human pregnancy lasts about forty weeks, the **embryo** developing in the **uterus** after **implantation**. Finger-like structures (**villi**) grow from the embryo and develop into the **placenta**.

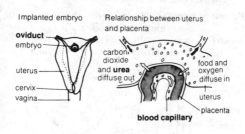

Implanted embryo

oviduct
embryo
uterus
cervix
vagina

Relationship between uterus and placenta

carbon dioxide and **urea** diffuse out

food and oxygen diffuse in

uterus

placenta

blood capillary

During pregnancy the **cells** of the embryo continually divide and differentiate, and the growing embryo (**foetus**) becomes suspended in a water sac, the **amnion**. The placenta extends into the *umbilical cord* which connects with the **abdomen** of the foetus.

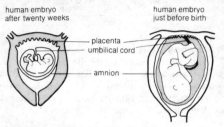

human embryo after twenty weeks

human embryo just before birth

placenta
umbilical cord
amnion

See **Birth; Fertilization in man**.

Primary growth (Apical growth) Increase in length and complexity of flowering plant **roots** and **shoots**, as the result of **cell division**, *cell elongation*, and **cell differentiation**, at the root and shoot *tips*. Such growing points are called **meristems**.

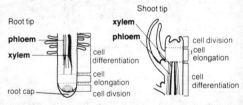

Root tip

phloem

xylem

root cap

cell differentiation
cell elongation
cell divsion

Shoot tip

xylem

phloem

cell division
cell elongation
cell differentiation

Progeny The offspring of **reproduction**.

Progesterone Hormone secreted by mammalian **ovaries**, which prepares the **uterus** for **implantation** and prevents further **ovulation** during **pregnancy**.

Protease Any **enzyme** which breaks down **protein** into **peptides** or **amino acids**, by **hydrolysis**, for example, **pepsin**, **trypsin**.

Proteins Organic **compounds** containing the elements *carbon*, *hydrogen*, *oxygen* and *nitrogen* and consisting of long chains of sub-units called **amino acids**.

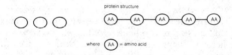

protein structure

where (AA) = amino acid

These chains may then be combined with others, and folded in several different ways, with various types of chemical bonding between chains and parts of chains, giving very large and complex molecules.

Proteins are the 'building blocks' of **cells** and **tissues** being important constituents of **muscle**, **skin**, **bone**, etc. Proteins also play a vital role as **enzymes** while some **hormones** are protein in structure. See **Amino acids; Peptide**.

Protoplasm All the material within and including the *cell membrane* of a cell, i.e., protoplasm consists of **nucleus** *and* **cytoplasm**.

Pulmonary artery See **Pulmonary vessels**; **Heart**; **Circulatory system**.

Pulmonary vein See **Pulmonary vessels**; **Heart**; **Circulatory system**.

Pulmonary vessels In mammals, **blood vessels**, which because of their special functions, disagree with the general rule that **arteries** carry *oxygenated blood*, and **veins** carry *deoxygenated blood*. The *pulmonary artery* carries *deoxygenated blood* from the right **ventricle** to the **lungs**, and the *pulmonary vein* carries *oxygenated blood* from the lungs to the left **atrium**. See **Heart**; **Circulatory system**.

Pulse rate Regular beating in **arteries** due to rhythmic movement of **blood** resulting from **heart beat**. Pulse rate can be detected in the human body where an artery is close to the skin surface, for example, at the wrist. In an adult human, pulse rate varies from about seventy beats per minute at rest, to over one hundred beats per minute during exercise.

Punnet square Graphic method used in **genetics** to calculate the results of all possible **fertilizations** and hence the **genotypes** and **phenotypes** of **progeny**. In a punnet square, the symbols used to represent one of the parent's **gamete** genotypes is written along the top and that of the other parent down the side. The permutations possible during fertilization are worked out by matching male and female gametes. See **Monohybrid inheritance**; **Backcross**; **Incomplete dominance**.

Pupa Stage in the **life history** of some insects between **larva** and **imago**, during which a radical change in form occurs. See **Metamorphosis**.

Pure-breeding Describing an inherited trait controlled by a **homozygous** pair of alleles, and which in successive *self-crosses* reappears generation after generation. See **Monohybrid inheritance**.

Pyramid of numbers Diagram illustrating the relationship between members of a **food chain**, showing that the organisms at the end of the chain are usually *fewer in number*, and *larger in size*. For example:

The decrease in numbers is the more significant, being caused by **energy** losses at each link in the chain, i.e., each organism in a food chain uses up energy in various activities such as heat production and movement. This energy is lost to the subsequent organisms in the chain and as a consequence, the reduced energy can only support a smaller number of individuals.

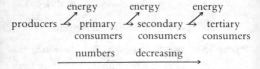

Quadrat A square of vegetation, randomly chosen to study the distribution of **species** in an area. The usual size is one metre square.

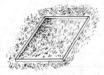

Radicle Embryonic **root** of **seed** plants which is the first structure to emerge from the seed during **germination**. See **Seed**; **Germination**.

Radius Anterior of the two **bones** of the lower region of the tetrapod forelimb. In man, the shorter of the two bones of the forearm. See **Endoskeleton**.

Receptor (sense organ)   Specialized tissue in an animal which detects **stimuli** from the **environment** and which, by sending **nerve impulses** through the **nervous system**, causes **responses** to be made. See **Sensitivity**.

Recessive   One of a pair of **alleles** which is only expressed in a **homozygous phenotype**. The converse of **dominant**. See **Monohybrid inheritance**, **Backcross**, **Incomplete dominance**.

Rectum   Terminal part of vertebrate **intestine** in which **faeces** are stored prior to expulsion via the **anus** or **cloaca**. See **Digestion**.

Red blood cell (Red blood corpuscle; Erythrocyte)   Most numerous **cell** of vertebrate **blood**, responsible for transporting *oxygen* from the **lungs** to the **tissues**. In man, they are made in **bone** marrow, and are biconcave discs, without **nuclei**.

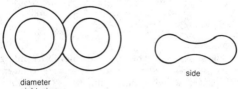

diameter
eight microns

side

Red blood cells contain **haemoglobin** which combines with oxygen as blood passes through the lungs, forming a compound called *oxyhaemoglobin*.

At the tissues, this unstable compound breaks down, thus releasing oxygen to the cells.

$$\text{haemoglobin} + \text{oxygen} \underset{\text{tissues}}{\overset{\text{lungs}}{\rightleftharpoons}} \text{oxyhaemoglobin}$$

Reflex action A rapid involuntary **response** to a **stimulus**, occurring in most animals and in vertebrates, mediated by the **spinal cord**. Reflex actions can be important in protecting animals from injury, for instance, in man, the withdrawal of the hand from a hot object.

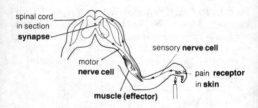

The structures involved and the **nerve impulses** responsible for reflex actions constitute a *reflex arc* which is set up when a nerve impulse is initiated at a receptor. The impulse is transmitted along a *sensory neurone* to the spinal cord where it crosses a **synapse** to a *motor neurone*. When a reflex arc operates, nerve impulses are also sent from the spinal cord to the

brain. Thus, although the response is initiated by the spinal cord, it is through the brain that the animal is aware of what has happened.

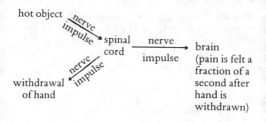

Regeneration The regrowth by an organism of **tissues** or **organs** which have been damaged or removed. Regeneration is common among plants and lower animals. For example:

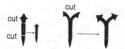

In higher animals, this degree of regeneration is not possible, due to the complexity of the **cells** and tissues present in such animals. Thus in mammals, *wound healing* is the only form of regeneration possible.

Reproduction The process by which a new organism is produced from one or a pair of parent organisms. See **Asexual reproducion**; **Sexual reproduction**.

Respiration The reactions by which organisms release the chemical **energy** of food, for example, **glucose**. The energy is used to synthesize **A.T.P.** from *A.D.P.* and is then available for other metabolic processes, for example, **muscle** action.

Aerobic respiration occurs in the *presence of oxygen* within the **mitochondria** of **cells**.

glucose + oxygen   →   carbon + water dioxide

$C_6H_{12}O_6$ + $6O_2$ $6CO_2$ + $6H_2O$

               A.D.P.   A.T.P.

Anaerobic respiration occurs in the *absence of oxygen* within the **cytoplasm** of **cells**, and provides a lower **A.T.P.** yield than aerobic respiration.

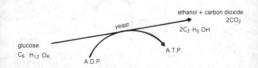

ethanol + carbon dioxide
2CO₂

yeast

2C₂ H₅ OH

glucose
C₆ H₁₂ O₆

A.T.P.

A.D.P.

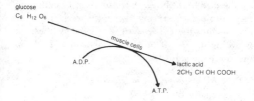

glucose
$C_6 H_{12} O_6$

muscle cells

A.D.P.

lactic acid
$2CH_3 CH OH COOH$

A.T.P.

Response Any change in an organism made in reaction to a **stimulus**. see **Sensitivity**.

Retina Light-sensitive **tissue** lining the interior of the vertebrate eye, and consisting of two types of **cells** (**rods** and **cones**). See **Eye**.

Rhizome Organ of **vegetative reproduction** in flowering plants consisting of a horizontal underground **stem** growing from a parent plant. The tip of the rhizome is a bud from which grows a new plant. Plants which have rhizomes include *iris* and many types of *grass*.

Grass rhizome

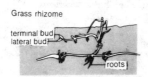

terminal bud
lateral bud

roots

R.N.A. (**Ribose Nucleic Acid**) Nucleic acid synthesized by **D.N.A.** in the **nucleus** of **cells**, and responsible for carrying the *genetic code* from the nucleus into the **cytoplasm** where the synthesis of **proteins** occurs. R.N.A. differs from D.N.A. in the following ways:

R.N.A. is a *single polynucleotide chain*.
The sugar group is *ribose*.
Thymine is replaced by *uracil*.

See **D.N.A.**; **Nucleic acids**.

Rod Light-sensitive **nerve cell** in the **retina** of the vertebrate **eye**. See **Eye**.

Root Part of a flowering plant that normally grows down into the **soil**. Its functions are (1) absorption of water and **mineral salts** from soil (2) to anchor the plant in the soil (3) in some plants, for example, turnip, storage of food.

Structure (**dicotyledon**)

Transverse section   Longitudinal section

epidermis
phloem
xylem
cortex
root hairs
root hair

Root cap Cap-shaped layer of **cells**, covering the apex of the growing *root tip*, and protecting it, as the root grows through the **soil**.

Longitudinal section through root

root tip — root cap
meristem

Root hairs Tubular projections from **root epidermis cells**, the nucleus usually passing into the hair. Root hairs enormously increase the *surface area* of the root, and are the principal absorbing **tissue** of the plant. Water enters root hairs from the **soil** by osmosis, while **mineral salts** are absorbed by **active transport**. The water and mineral salts then pass through the **cortex** cells and enter **xylem** vessels from where they are transported throughout the plant via the **transpiration stream**.

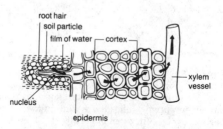

root hair
soil particle
film of water — cortex
nucleus
epidermis
xylem vessel

Root nodules   Swellings on the roots of *legumi-nous plants* (for example, clover, bean, pea). Root nodules contain bacteria of the **genus** *Rhizobium* which convert the *nitrogen* of **soil** air into **organic** nitrogen compounds which can be used by the legumes. This is called **nitrogen fixation**. See **Nitrogen cycle**.

Roughage   Important component of human **bal-anced diet** consisting mainly of the **cellulose** in plant **cell** walls. Although indigestible by man, roughage adds bulk to food and enables the **muscles** of the **alimentary canal** to grip the food and keep it moving by **peristalsis**.

Saliva   Fluid secreted by *salivary glands* into the **mouths** of many animals in order to moisten and lubricate food. In some mammals, including man saliva contains the **enzyme** *salivary amylase (ptyalin)* which digests **starch** into *maltose*.

Saprophyte   An organism that feeds on dead and decaying plants and animals, causing decomposi-tion. Many *fungi* and **bacteria** are saprophytic and play an important role in recycling nutrients. See **Carbon cycle**; **Nitrogen cycle**.

Scapula Dorsal part of tetrapod shoulder-girdle. In man, the *shoulder blade*. See **Endoskeleton**.

Scientific method The procedures by which scientific investigations should be made. Scientific method involves the following steps.

1) **Observation** An occurrence is seen to happen on more than one occasion. For example, **starch** in plant **seeds** apparently supplies **energy** during **germination**.

2) **Problem** The observation is questioned; for example, how does starch which is a long chain **carbohydrate** become suitable as a *respiratory substrate*?

3) **Hypothesis** The suggestion of a possible solution, for example, an **amylase enzyme** within seeds degrades starch to **glucose**.

4) **Experiment** Test the hypothesis, for example, add seed extract to starch, and test for glucose.

5) **Theory** The proposal of a solution to the problem based on experimental evidence, for example, in plant seeds, an amylase enzyme degrades starch to glucose which then acts as a respiratory substrate during germination.

All valid scientific investigations follow the guidelines of the scientific method and must include **control experiments** which are identical to the test experiment *in all aspects except one*. The control provides a standard with which the test experiment can be compared, by showing that any change occurring in the test experiment was due to the factor missing from the control and would not have happened anyway. For example, in the seed/starch experiment, a tube containing starch alone would be a suitable control.

Sclerotic   External protective layer of vertebrate eye-ball. See **Eye**.

Secondary growth (Secondary thickening) Increase in girth which occurs in woody flowering plants each year due to the activity of the **meristem** called **cambium**, which lies between **xylem** and **phloem**.

Secondary growth in a woody stem

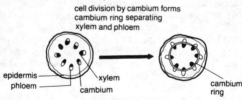

Further cambium activity produces new xylem and phloem cells, while epidermis becomes bark. Each year a new ring of xylem is added, increasing the girth of the plant and pushing the phloem and cambium ring outwards, the central core of *pith* cells being squashed out of existence. The annual addition of xylem (wood) is called an *annual ring*.

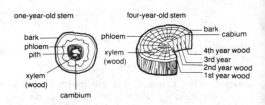

one-year-old stem

four-year-old stem

Secondary sexual characteristics Features which distinguish between adult male and female animals (excluding **gonads** and associated structures). The development of such features, for example, lions mane. stag's antlers, and in humans, breast development in females, facial hair in males, etc., is usually controlled by sex **hormones**.

Seed The structure that develops from an **ovule** after **fertilization** in flowering plants, and which grows into a new plant. Seeds are enclosed within a **fruit**. Within the seed, the **embryo** becomes differentiated into an embryonic **shoot** bud (**plumule**) and **root** (**radicle**) and either one or two seed leaves (**cotyledons**).

Section through bean seed

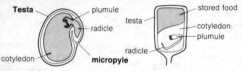

See **Fruit and seed dispersal**; **Germination**.

Selectively-permeable membrane   A porous surface which allows particles to pass on the basis of size, i.e., smaller particles pass easily, while larger particles are restricted. **Cell** membranes are selectively permeable or alternatively *semi-permeable*. See **Cell**, **Osmosis**.

Semi-circular canals   Tubes within the vertebrate *inner ear*, which are important in maintaining *balance*. See **Ear**.

Sense organ   See **Receptor**; **Sensitivity**.

Sensitivity (**Irritability**)   The ability of living organisms to respond to changes in **environmental stimuli**, such as *heat*, *light*, *sound*, etc. Sensitivity enables organisms to be aware of changes in their environment and thus they can make appropriate **responses** to any changes that may occur. Certain parts of animals, for example, **eyes**, **ears**, **skin**, are sensitive to particular environmental stimuli and are called **sense organs** or **receptors**. Similarly, plant **tissues** such as *shoot tips*, are receptors, being important in **tropisms**.

Due to stimuli from the environment, responses are initiated in specialized structures called **effectors**, for example, **muscles**. The responses made by an organism constitute its *behaviour*.

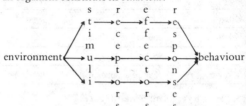

Sensitivity in mammals

In mammals the **receptors** are specialized **cells** connected to the **brain** or **spinal cord** which together make up the **central nervous system** (C.N.S.). In **response** to a **stimulus**, the receptors initiate a **nerve impulse** which is transmitted by **nerve cells** to the C.N.S., i.e., the receptor converts the **energy** of the stimulus into the *electrical energy* of the nerve impulse.

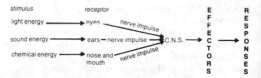

Summary

Sense	Stimulus	Receptor
smell	chemicals	nose
taste	chemicals	mouth
touch	contact	skin
hearing	sound	ears
sight	light	eyes
balance	change of position	inner ear

See **Ear; Eye; Skin; Smell; Taste**.

Sex chromosome Any chromosome that is involved in **sex determination**. In **diploid** human **cells** there are 46 chromosomes made up of 23 **homologous** pairs, one of the pairs being described as *sex chromosomes*. In the *female* the two sex chromosomes are similar and are called *X chromosomes*. The female **genotype** is thus XX (*homogametic*). In the male, one of the pair is distinctly smaller, and is called the Y *chromosome*. The male genotype is thus XY (*heterogametic*).

The male is not always the heterogametic sex. For example, in *birds*, the male is XX, and the female is XY, while in some *insects*, the female is XX and the male is XO, the Y chromosome being absent.

The sex chromosomes, as well as determining sex, also contain **genes** controlling other traits, resulting in what is known as **sex linkage**.

Sex determination The method by which the sex of a **zygote**, is determined, the most common method being by sex **chromosomes**. Consider the **genotypes** of a *human* male and female:

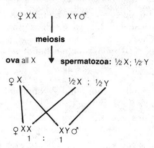

That is:

1) A Y-bearing sperm may fertilize an ovum giving a zygote genotype XY and **phenotype** *male*.

2) An X-bearing sperm may fertilize an ovum giving a zygote genotype XX and phenotype *female*.

Since *half* the sperms are X, and *half* are Y, there is an *equal* chance of the zygote being male or female.

Sex linkage The presence of **genes**, unconnected with sexuality, on a **sex chromosome**, resulting in certain traits appearing in only one sex. In humans, sex-linked genes are carried only on *X chromosomes*, the *Y chromosome* being concerned entirely with sexuality.

Example:

The gene for *colour blindness* in humans is carried on the *X chromosome*. *Normal vision* is **dominant** to *colour blindness*.

If N = normal and n = colour blind:

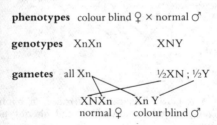

phenotypes colour blind ♀ × normal ♂

genotypes XnXn   XNY

gametes all Xn   ½XN ; ½Y

XNXn   Xn Y
normal ♀   colour blind ♂

In this case, the **heterozygous** ♀ XNXn is called a '*carrier*' since she has normal vision but carries the **recessive allele**. Thus if crossed with a normal male:

	carrier ♀	×	normal ♂
XNXn			XNY
gametes ½XN ; ½XN			½XN ; ½Y

sperms

		XN	Y
o	XN	XNXN	XNY
v			
a	Xn	XNXn	XnY

p u n n e t

s q u a r e

Progeny genotypes: XNXN   XNXn
phenotypes: normal ♀   carrier ♀

           genotypes: XNY   XnY
phenotypes: normal ♂   colour blind ♂

That is, there is a possibility that *half* the sons will be colour blind, and *half* the daughters will be carriers.

A more serious sex-linked trait is *haemophilia* (prolonged bleeding) but its transmission and inheritance is the same as above.

Sexual reproduction Reproduction involving the joining or fusing of two *sex cells* (**gametes**) one from a male parent, and one from a female parent. Gametes are **haploid** and when they fuse (**fertilization**), the resulting composite cell (**zygote**) has the **diploid** number of **chromosomes**. After fertilization, the zygote divides repeatedly, ultimately resulting in a new organism. Thus in humans:

Unlike **asexual reproduction**, the offspring of sexual reproduction are genetically unique (except for *identical* twins) because they obtain half their chromosomes from their male parent and half from their female parent. Thus each fertilization produces a new combination of chromosomes which in turn produce a new organism.

Shoot That part of a flowering plant which is above **soil** level, for example, **stem**, **leaves**, *buds*, **flowers**.

Short sight (Myopia) Human eye defect, mainly caused by the distance from **lens** to **retina** being longer than normal. This results in *distant* objects being focussed in front of the retina, giving blurred vision. Short sight is corrected by wearing *diverging (concave)* lenses.

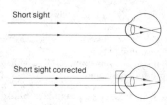

Skeleton The hard framework of an animal which supports and protects the internal organs. See **Endoskeleton; Exoskeleton**.

Skin Layer of **epithelial cells, connective tissue** and associated structures, that covers most of the body of vertebrates.

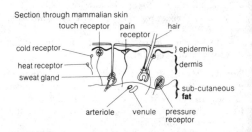

Section through mammalian skin

Mammalian skin consists of *two* main layers:

Epidermis is the outer layer consisting of:

1) *Cornified layer* Dead cells forming a tough protective outer coat.
2) *Granular layer* Living cells which ultimately form the cornified layer.
3) *Malpighian layer* Actively dividing cells which produce new epidermis.

Dermis is a thicker layer containing: **blood capillaries**, *hair follicles*, *sweat glands*, **receptor** cells sensitive to *touch*, *heat*, *cold*, *pain*, *pressure*.

Beneath the dermis, there is a layer of **fat** storage cells which also act as heat insulation.

Functions of mammalian skin

1) Protects against injury and **micro-organism** entry.
2) Reduces water loss by evaporation.
3) Acts as **receptor** for certain **environmental stimuli**.
4) In **homiothermic** animals, it is important in **temperature regulation**.

Small intestine Anterior region of vertebrate **intestine**. In man it consists of the **duodenum (about thirty centimetres, in length) and the ileum** (about seven metres in length). The duodenum receives food from the **stomach**. See **Digestion**.

Smell   The ability of animals to detect odours. In man, the **receptor cells** involved are in the nasal cavity, and are sensitive to chemical **stimuli. See Sensitivity**.

Soil   The *weathered* layer of the earth's crust intermingled with living organisms and the products of their decay.
The components of soil

(1) **Inorganic** particles (weathered rocks) (2) Water (3) **Humus** (4) Air (5) **Mineral salts** (6) **Micro-organisms** (7) Other organisms (for example, earthworms)
The importance of soil

   1) It is a **habitat** for a wide variety of organisms.
   2) It provides plants with water and mineral salts.
   3) Decomposition of dead organisms in soil releases minerals which can be used by other living organisms.

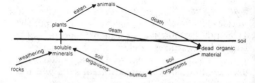

Soil depletion   The loss of **mineral salts** from soil when a crop is harvested, which like **soil**

erosion may render the soil infertile. Soil depletion can be prevented by, (1) **crop rotation**, (2) addition of **fertilizers**.

Soil erosion   The loss of **mineral salts** from **soil** due to the agricultural practices associated with crop growing, for example, repeated ploughing, deforestation, etc., which make the mineral-rich top soil less stable and more vulnerable to the effects of wind and rain. See **Soil depletion**.

Soil texture   The types and proportions of **soil inorganic** particles, of which four types are recognized based on size.

clay   silt   sand   gravel

increasing particle size

Soil texture has important effects on soil properties such as water retention and aeration. See **Soil types**.

Soil types   Numerous types of soil exist, but a simple classification recognizes three distinct types:
1) **Sandy (light) soil**   has a high proportion of the larger **inorganic** particles, and hence larger air spaces. Thus, sandy soil is well aerated and has good drainage but tends to lose **mineral salts** which are washed downwards (leeching).
2) **Clay (heavy) soil**   has a high proportion of small particles, which means that it retains water and

minerals, but is poorly aerated and can become waterlogged.

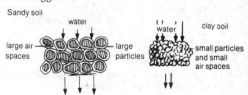

Sandy soil

water

large air spaces — large particles

clay soil

water

small particles and small air spaces

3) Loam soil is the most fertile soil, consisting of a balance of particle types and a good **humus** content. Soils of this type are well-aerated and drain freely, but still retain water and minerals.

Solution The mixture (usually a liquid) formed when one substance (the *solute*) dissolves in another (the *solvent*).

i.e. solute + solvent → solution
e.g sugar + water → sugar solution

Species Unit used in the **classification** of living organisms, describing any group which share the same general physical characteristics and which *can mate and produce fertile offspring*, for example, all dogs, despite variation in shape, size etc., are of the same species, but horses and donkeys are separate species within the same **genus**.

187

Spermatozoon (sperm) Small motile male **gamete** formed in animal **testes**, and usually having a **flagellum**. Sperms are released from the male in order to fertilize the female gamete. See **Fertilization; Meiosis**.

Sphincter Ring of **muscle** around tubular **organs**, which by contracting, can narrow or close the passage within the organ. For example, *anal sphincter*, *pyloric sphincter*. See **Digestion**.

Spinal cord That part of the vertebrate **central nervous system** which is enclosed within and protected by the backbone. It is a cylindrical mass of **nerve cells** which connect with the **brain** and also with other parts of the body via *spinal nerves*. The spinal cord consists of three regions:

1) An inner layer of *grey matter* consisting of **neurone** *cell-bodies*.
2) An outer layer of *white matter* consisting of *nerve fibres* running the length of the cord.
3) A fluid-filled central canal.

section through spinal cord

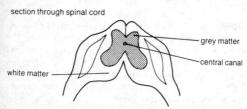

grey matter

central canal

white matter

The spinal cord conducts **nerve impulses** to and from the brain and is also involved in **reflex actions**.

Spiracle   One of many pores in the **cuticle** of insects, connecting the **tracheae** with the atmosphere. See **Gas exchange (insects)**.

Spleen   Organ in the **abdomen**, near the **stomach**, in most vertebrates. It produces **white blood cells**, destroys worn out **red blood cells**, and filters foreign bodies from the **blood**.

Spongy mesophyll   Tissue in a **leaf** situated between the **palisade mesophyll** and the lower **epidermis**. Spongy mesophyll **cells** are loosely packed, being separated by *air spaces* which allow **gas exchange** between the leaf and the atmosphere via the **stomata**. See **Leaf**.

Spore   Reproductive unit, usually microscopic, consisting of one or several **cells**, which becomes detached from a parent organism and ultimately gives rise to a new individual. Spores are involved in both **asexual** and **sexual reproduction** (as **gametes**) and are produced by certain *plants*, *fungi*, *bacteria* and *protozoa*. Some spores form a resistant resting stage of a **life history** while others allow rapid colonization of new **habitats**.

For example spore release in the bread mould Mucor

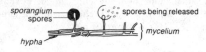

Stamen   Male part of a **flower** in which **pollen** grains are produced. Each stamen consists of a stalk (*filament*) bearing an **anther**.

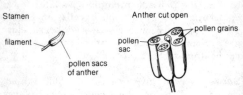

Starch   **Polysaccharide carbohydrate** consisting of chains of **glucose** units and important as an **energy** store in plants. Starch is synthesized during **photosynthesis** and is readily converted to glucose by **amylase enzymes**. See **Polysaccharides**.

Stem That part of a flowering plant that bears the *buds,* **leaves**, and **flowers**. Its *functions* are: (1) transport of water, **mineral salts** and **carbohydrate** (2) to raise the leaves above the **soil** for maximum air and light (3) to raise the **flowers**, and thus aid **pollination** (4) in *green* stems, **photosynthesis**.

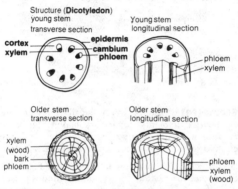

Structure (**Dicotyledon**)
young stem
transverse section

Young stem
longitudinal section

cortex
xylem

epidermis
cambium
phloem

phloem
xylem

Older stem
transverse section

Older stem
longitudinal section

xylem
(wood)
bark
phloem

phloem
xylem
(wood)

Sternum **Bone** in the middle of the ventral side of the **thorax** of tetrapods, to which most of the ventral ribs are attached, i.e., the *breast-bone*. See **Endoskeleton**.

Stimulus Any change in the **environment** of an organism which may provoke a **response** in the organism. See **Sensitivity**.

Stolon Organ of **vegetative reproduction** in flowering plants consisting of a horizontal **stem** growing from a bud on the parent organisms stem. Stolons grow above the soil and eventually the tip becomes established in the soil and develops into an independent plant. An example of a stolon is a *strawberry runner*, illustrated below.

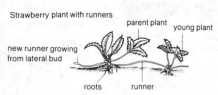

Strawberry plant with runners

parent plant

young plant

new runner growing from lateral bud

roots

runner

Stoma(ta) One of many small pores in the **epidermis** of plants, particulary **leaves**. The evaporation of water during **transpiration**, and **gas exchange**, occur via the stomata. See **Guard cells**.

Stomach Muscular sac in the anterior region of the **alimentary canal**. In vertebrates, food is passed to the stomach by **peristalsis** via the **oesophagus**.

In the stomach, food is mechanically churned by the peristaltic action of the walls and **protein digestion** is initiated. In **herbivores**, the stomach has several chambers for **cellulose** digestion.

From the stomach, food is passed into the **small intestine** through the pyloric **sphincter**. See **Pepsin**.

Surface area/volume ratio   The ratio

$$\frac{\text{surface area}}{\text{volume}} \quad \text{or} \quad \frac{\text{surface area}}{\text{weight}}$$

is significant to living organisms in several ways. It is diffficult to measure the surface area and volume of a plant or animal, but by using cubes as model organisms, the importance of the ratio can be seen.

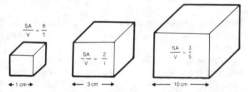

$\frac{SA}{V} = \frac{6}{1}$ ← 1 cm →

$\frac{SA}{V} = \frac{2}{1}$ ← 3 cm →

$\frac{SA}{V} = \frac{3}{5}$ ← 10 cm →

In the models, as the object becomes larger, its surface area becomes smaller *relative* to its volume. This is also true of living organisms and has special significance in terms of heat and water loss.

Surface area/volume, and heat loss   Heat is lost more rapidly from small animals because their *relatively* larger surface area allows easier heat loss to the air with the following consequences:

1) Small mammals such as mice eat relatively more food than larger mammals in order to generate **energy** to replace their high heat losses.
2) Very small birds and mammals are restricted to warm climates.

3) Birds and mammals in cold **habitats** are usually larger than the same **species** living in warm climates.

Surface area/volume, and water loss

Relative to their volume, small organisms have a larger evaporating surface and thus a greater tendency to lose water. This is important since many animals and plants have problems of excessive water loss, and this is obviously more serious the smaller the organism is.

Suspensory ligaments Structures holding the **lens** in place in the vertebrate **Eye**. See **Eye**, **Accommodation**.

Symbiosis A relationship between organisms of different **species** for the purpose of nutrition. Examples of symbiosis include **parasitism**, **mutualism** and **commensalism**, although the term is sometimes restricted to **mutualism**.

Synapse A microscopic gap between the **axon** of one **nerve cell** and the **dendrites** of another, across which a **nerve impulse** must pass. Nerve impulses arriving at a synapse cause **diffusion** of a chemical substance which crosses the gap and initiates nerve impulses in the next nerve cell.

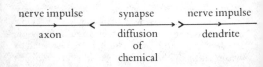

| nerve impulse | synapse | nerve impulse |
| axon | diffusion of chemical | dendrite |

Synovial membrane Membrane of **connective tissue** lining the capsule of a vertebrate moveable **joint**, being attached to the **bones** at either side of the joint. Synovial membrane secretes *synovial fluid* which bathes the joint cavity, lubricating the joint when the bones move and cushioning against jarring. See **Joint**.

Systole See **Heart beat**.

Taste The ability of animals to detect flavours. In man the **receptor cells** involved are *taste buds* which are sensitive to chemical **stimuli**, and are restricted to the **mouth**, particularly the *tongue*. There are four types of taste bud, sensitive to *sweet*, *sour*, *salt*, and *bitter*.

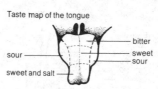

Taste map of the tongue

bitter
sour
sweet
sour
sweet and salt

See **Sensitivity**.

Taxis A locomotory movement of a simple organism or a **cell** in response to an **environmental stimulus**, for example, light. Such movements show a relationship to the direction of the stimulus, the movement either being towards (*positive*) or away (*negative*) from the source of the stimulus.

Taxes are named by adding a prefix which refers to the stimulus. Thus a taxis relative to light is a *phototaxis*. Examples:

> *Paramecium* is *negatively geotactic*, i.e., it swims away from gravity.
> Fruit flies are *positively phototactic*, i.e., they move towards light.
> Many **spermatozoa** are *positively chemotactic*, i.e., they move towards chemical substances released by **ova**.

Teeth Structures within the **mouth** of vertebrates, used for biting, tearing, and crushing food before swallowing.

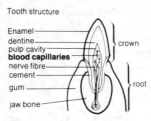

Tooth structure

Enamel
dentine
pulp cavity
blood capillaries
nerve fibre
cement
gum
jaw bone

crown

root

Enamel Hard substance covering the exposed surface of the tooth (the *crown*). It contains *calcium phosphate* and provides an efficient biting surface.
Dentine Substance similar to **bone**, forming the inner part of the tooth.

Pulp Soft **tissue** in the centre of the tooth containing **blood capillaries** which supply food and oxygen and nerve fibres which register pain if the tooth is damaged.

Root That part of the tooth, within the *gum*, and embedded in the *jaw-bone* by a substance called *cement*.

Types and functions of teeth in mammals

Incisors Chisel-shaped cutting teeth at the front of the mouth used for biting off pieces of food.

Canines Sharp-pointed tearing teeth near the front of the mouth, used for killing prey, and ripping off pieces of food.

Pre-molars and molars (cheek teeth) Broad crowned grinding teeth at the sides and back of the mouth, used for crushing food prior to swallowing. See **Dental formula**, **Carnivore**, **Herbivore**.

Temperature regulation

In **homiothermic** animals the mechanisms involved in maintaining body temperature within a narrow range (for example, in man, close to 37°C) so that the normal reactions of **metabolism** can take place. Some of the temperature regualtion methods employed by birds and mammals are outlined below:

1) Sub-cutaneous **fat** acts as an insulator.
2) Hair in mammals and feathers in birds trap air which is a good insulator.
3) In mammals evaporation of *sweat* from the **skin** surface has a cooling effect.

4) Superficial **blood vessels** *constrict* (*vasoconstriction*) in response to cold, diverting blood away from the skin surface, and thus reducing heat loss.

5) Superficial blood vessels *dilate* (*vasodilation*) in response to heat, bringing blood to the skin surface, from which heat can be lost to the atmosphere.

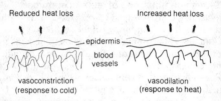

Reduced heat loss   Increased heat loss

epidermis
blood vessels

vasoconstriction   vasodilation
(response to cold)   (response to heat)

Tendon Band of **connective tissue** by which **muscles** are attached to **bones**.

Testa Outer protective coat of a **seed** formed from the **integuments** of the **ovule**, after fertilization. The testa is usually hard and dry and protects the seed from **micro-organisms** and insects.

Testis Principal reproductive organ in male animals which produces **sperms**. In vertebrates, the paired testes also produce sex **hormones**. See **Fertilization in man**.

Thorax In vertebrates, the part of the body containing **heart** and **lungs** (chest cavity). In mammals it is separated from the **abdomen** by the

diaphragm. In insects, the part of the body, anterior to the abdomen. See **Abdomen**.

Thyroid gland Endocrine gland in the neck region of vertebrates. When stimulated by T.S.H. from the *pituitary gland* it produces the **hormone** *thyroxine* which controls the rate of **growth** and development in young animals. For example in tadpoles thyroxine stimulates **metamorphosis**. See **Hormones**.

Tibia 1) One of the segments of the insect leg.

2) Anterior of the two **bones** in the lower hindlimb of tetrapods. In man, the *shin-bone*. See **Endoskeleton**.

Tissue In **multicellular** organisms, a group of similar **cells** specialized to perform a specific function. For example, **muscle**, **xylem**.

Tissue fluid See **Lymph**.

Trachea 1) In land vertebrates, the windpipe leading from the **larynx** and carrying air to the **lungs** where it divides into the **bronchi**. The trachea is supported by **cartilage** rings and has a ciliated **epithelium** that secretes **mucus** which traps dust and **micro-organisms**. See **Lungs**.
2) In insects one of a branching system of air tubes through which air diffuses into the **tissues** via the **spiracles**. See **Gas exchange (insects)**.

Transect A line marked off in an area, to study the types of **species** in that area, by sampling the organisms at different points along the line. Measurements of environmental factors, for example, *light*, *soil pH*, etc., may also be made along the line to discover any relationship between the distribution of particular species and these factors.

Translocation The transport and circulation of materials within plants. That is:

a) of water and **mineral salts** in *xylem* vessels via the **transpiration stream**.
b) of **carbohydrate** produced by **photosynthesis** and conducted through the plant in **phloem** sieve tubes.

Transpiration The evaporation of water vapour from plant **leaves** via the **stomata**.

Transpiration rate **Transpiration** is affected by several environmental factors:
1 Temperature Increased temperature increases water evaporation and thus increases transpiration.
2 Humidity (water content of air) Increased humidity causes the atmosphere to become saturated with water, thus reducing transpiration.
3 Wind Increased air movements increase transpiration by preventing the atmosphere around **stomata** from becoming saturated with water.

Thus transpiration rate will be greatest in *warm*, *dry*, *windy* conditions. If the rate of water loss by transpiration exceeds the rate of water uptake, **wilting** may occur.

Transpiration stream The flow of water through a plant resulting from **transpiration**. Water evaporates through the **stomata**, causing more water to be drawn by **osmosis** from adjacent **leaf cells (spongy mesophyll cells)**. The osmotic forces thus set up eventually cause water to be withdrawn from **xylem** vessels in the leaf, resulting in water being pulled through the xylem vessels from the **stem** and **roots**, i.e., water evaporation from leaves causes the flow of water (and **mineral salts**) throughout the plant.

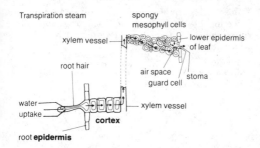

201

Tropism Plant **growth** movement in response to a **stimulus**, for example, light. Such movements are related to the *direction* of the stimulus, the plant *organ* involved growing either towards or away from it. Tropisms are named by adding a prefix which refers to the stimulus, for example, a light tropism is called a **phototropism**. Growth towards light is *positive phototropism*, while growth away is *negative phototropism*.

Tropisms are important because they cause plants to grow in such a way that they obtain maximum benefit from the **environment**, in terms of water, light, etc. Tropisms are caused by a plant **hormone** or **auxin** which accelerates growth by stimulating **cell division** and *elongation*. Uneven distribution of auxin causes uneven growth and leads to bending.

Trypsin Protease **enzyme** secreted by the vertebrate **pancreas**. See **Duodenum**.

Tuber Organ of **vegetative reproduction** in flowering plants. Tubers can form from **stems** or **roots** and consists of a food store and buds from which develop new plants.

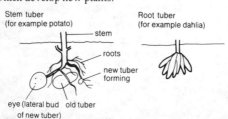

Stem tuber
(for example potato)

— stem

— roots

— new tuber forming

eye (lateral bud of new tuber)   old tuber

Root tuber
(for example dahlia)

Turgor   State of a plant **cell** after maximum water absorption. Water surrounding a cell enters by osmosis causing the **vacuole** to expand, pushing the **cytoplasm** against the cell wall and making the plant cell solid and strong. Such cells are said to be *turgid*.

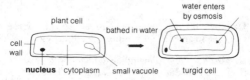

Turgid cells are important in supporting plants, conferring strength and shape. Young plants depend completely on turgor for support, although in older plants, support is obtained from **wood** formation.

Tympanum   (tympanic membrane)   Thin membrane separating the *outer* and *middle ear* in tetrapods, i.e., the *ear-drum*. See **Ear**.

Ulna   Posterior of the two **bones** of the lower region of the tetrapod forelimb. In man, the larger of the two bones of the forearm. See **Endoskeleton**.

Unicellular (organism)   An organism consisting of *one* **cell** only. Unicellular organisms include *protozoans*, **bacteria** and some *algae*. See **Multicellular**.

Urea Main *nitrogenous* excretory product of mammals. Urea is produced in the **liver** from the **deamination** of excess **amino acids** and then excreted by the **kidneys**.

$$H_2N- \underset{\underset{O}{\overset{\|}{}}}{C} - NH_2 \quad \text{urea}$$

Ureter In vertebrates, the tube carrying **urine** from the **kidney** to the **bladder**. See **Kidney**.

Urethra Tube in mammals which conveys **urine** from the **bladder** to the exterior. In male mammals it also serves as a channel for **spermatazoa** exit. See **Kidney**; **Fertilization in man**.

Uterus (womb) Muscular cavity in most female mammals that contains the **embryo(s)** during development. The uterus receives **ova** from the **oviduct** and connects to the exterior via the **vagina**. See **Fertilization in man**; **Pregnancy**.

Urine Solution of **urea** and *salts* in water produced by the mammalian **kidney**. It is stored in the **bladder** before discharge via the **urethra**.

Vacuole Fluid-filled space within **cell cytoplasm**, containing many compounds, for example, sugars in solution. Vacuoles are particularly important in maintaining **turgor** in plant cells. See **Cell**; **Contractile vacuole**.

Vagina Duct in most female mammals which receives the **penis** during **copulation**. It connects the **uterus** with the exterior and is the route by which the **foetus** is passed during **birth**. See **Fertilization in man**.

Valves Membranous structures within animal **circulatory systems** which allow **blood** to flow in one direction only.
Mitral valve (bicuspid valve) Two flaps between the left **atrium** and left **ventricle** of the **heart** in birds and mammals.
Tricuspid valve Three flaps between the *right atrium* and *right ventricle* of the mammalian heart.
Semi-lunar valves Half-moon shaped flaps in the mammalian heart between the *right ventricle* and *pulmonary artery*, and the *left ventricle* and **aorta**. Semi-lunar valves are also found in **lymphatics** and **veins**. See **Heart; Heart beat**.

Variation Differences in characteristics between members of the same **species**. There are two main types:

Continuous variations in which there are *degrees* of variation throughout the **population** showing **normal distribution** around a *mean*. For example, in humans: *height, weight, pulse rate.*

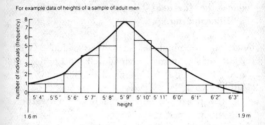

For example data of heights of a sample of adult men

Discontinuous variations are absolutely clear cut, i.e., there are no intermediate forms, for example, *blood groups in humans.* Discontinuous variations do not show normal distribution and are used when doing **genetics** crosses.

Variation within a species results either from *inherited* or **environmental** factors or a combination of both. Thus, a human being inherits **genes** influencing height for example, but will also be subject to environmental factors such as nutrition. Inherited variations are considered to be the basis of **evolution** by **natural selection.**

Vascular bundle Strand of longitudinal conducting **tissue** within plants consisting mainly of **xylem** and **phloem**. See **Root; Stem; Leaf**.

Vascular bundle in stem

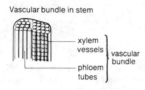

Vegetative reproduction (propagation) Asexual reproduction in plants by an outgrowth from a parent organism of a **multicellular** body which may become detached and develop independently into a new plant. See **Bulbs; Corm; Rhizome; Stolon; Tuber**.

Vein 1) One of the **vascular bundles** in a plant leaf.

2) **Blood vessel** which transports **blood** from the tissues to the **heart**. In mammals, veins carry *deoxygenated blood* (for an exception to this rule, see **Pulmonary vessels**) and form from smaller vessels called *venules* which carry blood from the **capillaries**. Veins are thin-walled, and since the **blood pressure** in veins is less than in **arteries**, they have

valves to prevent the blood flowing away from the heart.

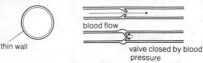

Section through a vein   Valve operation in veins

thin wall

blood flow

valve closed by blood pressure

Vena cava   The largest **vein** in the **circulatory system** of vertebrates. In mammals, either of the two main veins:

a) *superior vena cava* carries **blood** from the head, neck, and upper limbs into the *right* **atrium** of the **heart**.

b) *inferior vena cava* carries **blood** from the rest of the body, and lower limbs into the *right atrium*.

Ventricle   See **Heart; Heart beat**.

Vertebral column (backbone)   Series of closely arranged **bones** and/or **cartilages** (*vertebrae*) which runs dorsally from the skull to the tail in vertebrates. It is the principal longitudinal supporting structure and encloses and protects the **spinal cord**. See **Endoskeleton**.

Villi (singular **villus**) a) Finger-like projections in the vertebrate **intestine** where their large numbers increase the surface area available for **absorption** of food. See **Ileum**.

b) Finger-like projections which develop from the mammalian **placenta** into the **uterus** wall and thus increase the area of contact between maternal and embryonic **tissues**.

Virus Smallest known living particle having a diameter between 0.025 and 0.25 microns. Viruses are **parasites** infecting animals, plants and **bacteria**. Virus infections of man include *measles, polio, influenza*.

A virus particle consists of a **protein** coat surrounding a length of **nucleic acid**, either **D.N.A.** or **R.N.A.**

Virus infecting a bacterium

bacterium
virus becomes attached to bacterium

the virus nuclear material is injected into the bacterium and causes the assembly of new virus parts

the bacterium cell wall is ruptured, releasing many new viruses

Vitamins **Organic compounds** required in small quantities by living organisms. Like **enzymes**, vitamins play a vital role in chemical reactions within the body, often regulating an enzyme's action. Shortage of vitamins from the human diet leads to *deficiency diseases*. The properties of some important vitamins are summarized below.

Vitamin	Rich Sources	Effects of deficiency
Vitamin A	milk, liver, butter, fresh vegetables	night-blindness, retarded growth
Vitamin B_1	yeast, liver	*Beri-beri*: loss of appetite and weakness
Vitamin B_2	yeast, milk	*pellagra*: skin infections, weakness, mental illness
Vitamin C	citrus fruits, fresh geen vegetables	*scurvy*: bleeding gums, loose teeth, weakness
Vitamin D	eggs, cod liver oil	*rickets*: abnormal bone formation
Vitamin E	fresh green vegetables, milk	thought to affect reproductive ability
Vitamin K	fresh vegetables	blood clotting impaired

Vitreous humour Transparent jelly-like material which fills the cavity behind the **lens** of the vertebrate **eye**. See **Eye**.

Voluntary (striated) muscles Muscles connected to the mammalian **skeleton** and under the conscious control of the organism, for example, the limb muscles, muscles of face and mouth, etc. Voluntary muscles involved in limb movement are attached to **bones** by **tendons** and cause movement by contracting and thus pulling on bones, particularly at **joints**. See **Involuntary muscles, Antagonistic muscles**.

White blood cell (**White blood corpuscle, Leucocyte**) One of various types of **blood cell** found in most vertebrates. Their function is in defence against **micro-organism** infection, which they achieve by **phagocytosis** or by **antibody** production.

Wild type An organism having a **phenotype** or **genotype** which is characteristic of the majority of the **species** in natural conditions.

Wilting Plant condition occurring when water loss by **transpiration** exceeds water uptake. The **cells** lose **turgor** and the plant droops.

Wood See **Xylem, Secondary growth**.

Xylem Tissue within plants which conducts water and **mineral salts**, absorbed by **roots** from the **soil**, throughout the plant. Xylem tissue consists of *vessels* formed from columns of **cells**, in which the horizontal cross walls have disintegrated, the cell contents have died, and thus long continuous tubes are formed.

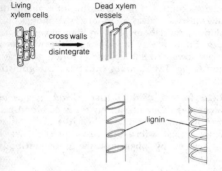

The vessels thus formed are strengthened by a compound called **lignin**, and ultimately form the *wood* of the plant. Associated with xylem vessels, and providing additional strength, are specialized cells called *xylem fibres*, some of which are useful, for example, *flax*. Thus xylem is commercially important as a source of wood and fibres. See **Leaf**, **Root**, **Secondary growth**, **Stem**.

Yolk A store of food material, mainly **protein** and **fat** present in the eggs (**ova**) of most animals. In fish, reptiles, and birds, the yolk is contained within a *yolk sac* which is absorbed into the **embryo** as the yolk is used.

Zygote The **diploid cell** resulting from the fusion of two **gametes** during **fertilization**. Normally the zygote undergoes *cleavage* by **mitosis** immediately after fertilization, resulting initially in a tiny ball of cells, which in mammals becomes embedded in the wall of the **uterus** (**implantation**).

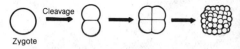

Zygote

See **Fertilization in man; Pregnancy**.